Proceedings

Ein stetig steigender Fundus an Informationen ist heute notwendig, um die immer komplexer werdende Technik heutiger Kraftfahrzeuge zu verstehen. Funktionen, Arbeitsweise, Komponenten und Systeme entwickeln sich rasant. In immer schnelleren Zyklen verbreitet sich aktuelles Wissen gerade aus Konferenzen, Tagungen und Symposien in die Fachwelt. Den raschen Zugriff auf diese Informationen bietet diese Reihe Proceedings, die sich zur Aufgabe gestellt hat, das zum Verständnis topaktueller Technik rund um das Automobil erforderliche spezielle Wissen in der Systematik aus Konferenzen und Tagungen zusammen zu stellen und als Buch in Springer.com wie auch elektronisch in Springer Link und Springer Professional bereit zu stellen. Die Reihe wendet sich an Fahrzeug- und Motoreningenieure sowie Studierende, die aktuelles Fachwissen im Zusammenhang mit Fragestellungen ihres Arbeitsfeldes suchen. Professoren und Dozenten an Universitäten und Hochschulen mit Schwerpunkt Kraftfahrzeug- und Motorentechnik finden hier die Zusammenstellung von Veranstaltungen, die sie selber nicht besuchen konnten. Gutachtern, Forschern und Entwicklungsingenieuren in der Automobil- und Zulieferindustrie sowie Dienstleistern können die Proceedings wertvolle Antworten auf topaktuelle Fragen geben.

Today, a steadily growing store of information is called for in order to understand the increasingly complex technologies used in modern automobiles. Functions, modes of operation, components and systems are rapidly evolving, while at the same time the latest expertise is disseminated directly from conferences, congresses and symposia to the professional world in ever-faster cycles. This series of proceedings offers rapid access to this information, gathering the specific knowledge needed to keep up with cutting-edge advances in automotive technologies, employing the same systematic approach used at conferences and congresses and presenting it in print (available at Springer.com) and electronic (at Springer Link and Springer Professional) formats. The series addresses the needs of automotive engineers, motor design engineers and students looking for the latest expertise in connection with key questions in their field, while professors and instructors working in the areas of automotive and motor design engineering will also find summaries of industry events they weren't able to attend. The proceedings also offer valuable answers to the topical questions that concern assessors, researchers and developmental engineers in the automotive and supplier industry, as well as service providers.

Peter E. Pfeffer
Editor

15th International Munich Chassis Symposium 2024

Volume 3: steering.tech and chassis.tech.

Springer Vieweg

Editor
Peter E. Pfeffer
Munich University of Applied Sciences
Munich, Germany

ISSN 2198-7432 ISSN 2198-7440 (electronic)
Proceedings
ISBN 978-3-662-71063-0 ISBN 978-3-662-71064-7 (eBook)
https://doi.org/10.1007/978-3-662-71064-7

This Springer Vieweg imprint is published by the registered company Springer-Verlag GmbH, DE, part of Springer Nature.
The registered company address is: Heidelberger Platz 3, 14197 Berlin, Germany

Prof. Dr. Peter E. Pfeffer
Hochschule München University of Applied Sciences
Scientific Director of the Symposium

Welcome

Current and future chassis and vehicle dynamics systems need to meet a variety of requirements. They should not only ensure safety through stability, braking ability, and controllability, but also enhance comfort at the same time. They are also expected to deliver optimum performance with the best possible driving experience, driving enjoyment, and handling. What is more, the systems should also be able to adapt to different driving and road conditions. At the same time, there is a demand for increased efficiency in order to reduce energy consumption and cut exhaust and noise emissions.
In an era of connected, software-based, and autonomous vehicles, support for connectivity and automation is crucial for optimum performance.
This wide range of topics sets the scene for the 15th International Munich Chassis Symposium chassis.tech plus. It brings together numerous experts in suspension systems, steering, brakes, tires/wheels, and automated driving for an exchange of ideas and experience.
Innovative research and development activities will be presented for the first time, forming the basis for constructive discussions. You can look forward to more than 50 lectures on a variety of topics, with keynotes from Dr. Ingo Albers (Porsche), Dr. Cyrille Roget (Michelin), Dr. Navid Samadi (Huawei), and Dr. Robert Fuchs (JTEKT) on the first day of the symposium, as well as an interactive discussion with the plenum. On the second day, Dr. Peter Holdmann (ZF) and Dr. Thomas Kersten (Volkswagen do Brasil) will present their keynote lectures.

We look forward to welcoming you to the Bayerischer Hof in the center of Munich or virtually via our live stream and hope you have an exciting and informative conference.

Stay at the cutting edge!

- Highly relevant technical papers presented by renowned speakers
- Networking in the international expert community
- Innovative products and services

Prof. Dr. Peter E. Pfeffer
Hochschule München
Wissenschaftliche Leitung des Symposiums

Herzlich willkommen

Für aktuelle und zukünftige Fahrwerke und Fahrdynamiksysteme gilt es, eine Vielzahl von Anforderungen zu erfüllen. So sollen sie nicht nur die Sicherheit durch Stabilität, Bremsfähigkeit und Kontrollierbarkeit gewährleisten, sondern auch zugleich den Komfort steigern. Darüber hinaus wird erwartet, dass sie eine optimale Performance bei einem Maximum an Fahrerlebnis, Fahrspaß und Handling liefern. Die Systeme sollen zudem adaptierfähig sein, um sich an unterschiedliche Fahr- und Straßenbedingungen anzupassen. Gleichzeitig wird eine erhöhte Effizienz gefordert, um den Energiebedarf sowie die Abgas- und Geräuschemissionen zu senken. In der Ära vernetzter, softwarebasierter und autonomer Fahrzeuge ist die Unterstützung von Konnektivität und Automatisierung für eine optimale Leistungsfähigkeit entscheidend.
Diese große Themenpalette spannt den Rahmen auf für das 15. Internationale Münchner Fahrwerk-Symposium chassis.tech plus. Es führt zahlreiche Fachleute für Radaufhängung, Lenkung, Bremsen und Reifen/Räder sowie für das automatisierte Fahren zum Erfahrungsaustausch zusammen.
Hier werden innovative Forschungs- und Entwicklungsaktivitäten erstmals vorgestellt und konstruktiv diskutiert. Freuen Sie sich auf über 50 vielfältige Vorträge mit Keynotes von Dr. Ingo Albers (Porsche), Dr. Cyrille Roget (Michelin), Dr. Navid Samadi (Huawei) und Dr. Robert Fuchs (JTEKT) am ersten Symposiumstag sowie auf eine Interaktive Diskussionsrunde mit dem Plenum. Am zweiten Tag werden Dr. Peter Holdmann (ZF) und Dr. Thomas Kersten (Volkswagen do Brasil) ihre Keynotes präsentieren.

Wir freuen uns, Sie im Bayerischen Hof im Herzen von München oder virtuell im Live-Stream begrüßen zu dürfen, und wünschen Ihnen eine anregende Veranstaltung.

Peter Pfeffer

Profitieren Sie von Ihrem Wissensvorsprung!

- Praxisrelevante Fachvorträge von hochkarätigen Referenten
- Netzwerken in der internationalen Experten-Community
- Innovative Produkte und Dienstleistungen

Conference Report

Chassis Trends from Tires to Steer-by-Wire

What's on the chassis community's mind? At chassis.tech plus 2024, the latest developments in the fields of suspension, steering, brakes and tires/wheels will be discussed.

Active chassis technology for sports cars, sustainable tires, driving dynamics for electric cars and steer-by-wire: The **15th International Munich Chassis Symposium** will once again bring the chassis community together in 2024 to exchange experiences. Alexander Heintzel, Editor-in-Chief of the ATZ | MTZ Group, and Prof. Dr. Peter E. Pfeffer from HM Hochschule München University of Applied Sciences will welcome 460 participants from all over the world, 400 of whom will be on site at the Hotel Bayerischer Hof in Munich.

As always, the event, which is also being held virtually, begins with the keynote speeches. **Dr. Ingo Albers**, VP Drive System at Porsche, will kick things off. He will present the Porsche Active Ride suspension, which is used in the revised Panamera and Taycan models. The air suspension with two-valve technology on each shock absorber improves the balance between driving dynamics and ride comfort. It has an electrically operated hydraulic pump that can be used to actively build up forces in the rebound or compression direction. The dampers operate at up to 13 Hz and keep the vehicle horizontal at all times. Body movements are reduced to a minimum. A Taycan is now also significantly faster on the racetrack thanks to the newly developed suspension. Functions such as the overcompensation of pitching and rolling tendencies or the lifting of the body when getting in and out of the car ensure greater comfort.

Reducing the CO_2 Footprint of Tires

Dr. Cyrille Roget from Michelin describes what makes an environmentally friendly tire. With a view to Euro 7 and the advancing electrification, which also involves special tires for heavier electric vehicles, he presents the method chosen by the tire manufacturer for life cycle assessment. It should make it possible to determine the most important levers for effectively reducing a tire footprint, regardless of the drive system. When manufacturing sustainable tires, it is important to find the right mix of recycled materials, efficiency, durability and reduction of particulate emissions. The tire manufacturer's life cycle assessment is based on the EU Commission's proposals. Over 80 % of the environmental impact of tires occurs during the use phase: Roget cites rolling resistance, durability and particle emissions as the most important criteria. In addition to the vulcanization process, which Michelin is converting to an electric system, the extraction of raw materials and the construction of the tire also play a decisive role in the sustainable production of tires.

The second keynote session will begin with **Dr. Navid Samadi**, Chief Engineer, Intelligent Chassis and Vehicle Performance at Huawei. In his presentation, he will provide an overview of the Chinese tech group's automotive activities. Huawei technology can be found in various Chinese cars, for example in the sporty Luxeed S7 electric sedan, which is manufactured by Chery. The Tesla Model S-sized vehicle combines a chassis with double wishbones, air suspension and semi-active damping with intelligent control. The 2.2 t electric vehicle with 365 kW (496 hp) from two electric motors uses the intelligent torque system "Adaptive Torque System" and "Dynamic Corner Control" in Sport+ mode to improve dynamic cornering behavior. The control system ensures an intelligent distribution of forces through the electric motors at the front and rear, increasing vehicle dynamics as well as vehicle safety thanks to the "Antiskid Mode".

Outlook on Steer-by-Wire

Dr. Robert Fuchs from JTEKT Corporation gives a brief overview of advances in steer-by-wire systems. The trends of automated driving and electrification are helping the development of steer-by-wire and vice versa. According to Fuchs, system reliability can be achieved through redundancy. One aim of steer-by-wire systems is to improve safety and the driving experience by allowing the vehicle to take over steering tasks and share them with the driver. This strengthens confidence in the technical systems and in turn helps to establish automated driving. "Shared control" works with "hands on/off" recognition and also enables the steering wheel to be completely locked and the driver to take over quickly. Concepts for steer-by-wire can also support the trend towards electric cars with fold-away steering wheels, as they offer more space in the interior when stationary during charging, for example.

Patrick Schäfer

Tagungsbericht

Fahrwerktrends vom Reifen bis Steer-by-Wire

Was beschäftigt die Fahrwerk-Community? Auf der chassis.tech plus 2024 werden die neuesten Entwicklungen in den Bereichen Radaufhängung, Lenkung, Bremsen und Reifen/Räder diskutiert.

Aktive Fahrwerkstechnologie für Sportwagen, nachhaltige Reifen, Fahrdynamik bei E-Autos und Steer-by-Wire: Das **15. Internationale Münchner Fahrwerk-Symposium** bringt auch 2024 die Fahrwerk-Community wieder zum Erfahrungsaustausch zusammen. Alexander Heintzel, Chefredakteur der ATZ | MTZ-Gruppe, und Prof. Dr. Peter E. Pfeffer von der Hochschule München begrüßen 460 Teilnehmer aus aller Welt, davon 400 vor Ort im Hotel Bayerischer Hof in München.

Wie immer beginnt das Event, das auch virtuell ausgerichtet wird, mit den Keynotes. Den Anfang macht **Dr. Ingo Albers**, Leiter Fahrsystem bei Porsche. Er stellt das Aktivfahrwerk „Porsche Active Ride" vor, das in den überarbeiteten Modellen Panamera und Taycan zum Einsatz kommt. Das Luftfahrwerk mit Zweiventiltechnik an jedem Stoßdämpfer verbessert die Balance zwischen Fahrdynamik und Fahrkomfort. Es verfügt über eine elektrisch betriebene Hydraulikpumpe, mit der aktiv Kräfte in Zug- oder Druckrichtung aufgebaut werden können. Die Dämpfer arbeiten mit bis zu 13 Hz und halten das Fahrzeug stets horizontal. Karosseriebewegungen werden auf ein Minimum reduziert. Ein Taycan sei nun auch aufgrund des neu entwickelten Fahrwerks deutlich schneller auf der Rennstrecke. Für mehr Komfort sorgen Funktionen wie die Überkompensation von Nick- und Wankneigungen oder das Anheben der Karosserie beim Ein- und Aussteigen.

CO_2-Fußabdruck von Reifen verringern

Was einen umweltfreundlichen Reifen ausmacht, schildert indes **Dr. Cyrille Roget** von Michelin. Er stellt mit Blick auf Euro 7 und der fortschreitenden Elektrifizierung, die auch spezielle Reifen für schwerere E-Fahrzeuge mit sich bringt, die vom Reifenhersteller gewählte Methode zur Ökobilanzierung vor. Sie soll es ermöglichen, die wichtigsten Hebel zur effektiven Reduzierung eines Reifen-Fußabdrucks zu bestimmen, egal für welchen Antrieb. Bei der Herstellung nachhaltiger Reifen gilt es, die richtige Mischung aus Recycling-materialien, Effizienz, Lebensdauer und Verringerung der Partikelemissionen zu finden. Bei seiner Ökobilanzierung orientiert sich der Reifenhersteller an den Vorschlägen der EU-Kommission. Über 80 % der Umweltauswirkungen von Reifen entstehen dabei während der Nutzungsphase: Rollwiderstand, Langlebigkeit und Partikelemissionen nennt Roget dabei als wichtigste Kriterien. Für die nachhaltige Produktion von Reifen spielen neben dem Vulkanisierungsprozess, den Michelin auf ein elektrisches System umstellt, auch die Gewinnung von Rohstoffen und die Konstruktion des Reifens eine entscheidende Rolle.

Die zweite Keynote-Session beginnt mit **Dr. Navid Samadi**, Chief Engineer, Intelligent Chassis and Vehicle Performance bei Huawei. Er gibt in seinem Vortrag einen Überblick über die Automotive-Aktivitäten des chinesischen Tech-Konzerns. Technik von Huawei steckt in verschiedenen chinesischen Autos, zum Beispiel in der sportlichen Elektrolimousine Luxeed S7, das von Chery hergestellt wird. Das Fahrzeug im Format eines Tesla Model S kombiniert ein Fahrwerk mit Doppelquerlenkern, Luftfederung und semiaktiver Dämpfung mit einer intelligenten Steuerung. Das 2,2 t schwere Elektrofahrzeug mit 365 kW (496 PS) aus zwei E-Motoren nutzt im Sport+-Mode das intelligente Drehmomentsystem „Adaptive Torque System“ und die „Dynamic Corner Control“, um das dynamische Kurvenverhalten zu verbessern. Die Steuerung sorgt für eine intelligente Verteilung der Kräfte durch die E-Motoren vorne und hinten, sodass sich die Fahrzeugdynamik ebenso erhöht wie die Fahrzeugsicherheit durch den „Antiskid Mode“.

Ausblick auf Steer-by-Wire

Dr. Robert Fuchs von JTEKT Corporation gibt einen kurzen Überblick über Fortschritte bei Steer-by-Wire-Systemen. Die Trends automatisiertes Fahren und Elektrifizierung helfen bei der Entwicklung von Steer-by-Wire und umgekehrt. Die Zuverlässigkeit der Systeme lässt sich laut Fuchs durch Redundanz herstellen. Ein Ziel von Steer-by-Wire-Systemen ist es, die Sicherheit und das Fahrgefühl zu verbessern, indem das Fahrzeug Lenkaufgaben übernimmt und sie mit dem Fahrer teilt. Das stärkt das Vertrauen in die technischen Systeme und hilft wiederum bei der Etablierung des automatisierten Fahrens. „Shared Control“ arbeitet mit einer „Hands on/off“-Erkennung und ermöglicht unter anderem auch die komplette Sperrung des Lenkrads sowie eine schnelle Übernahme durch den Fahrer. Konzepte für Steer-by-Wire können durch wegklappbare Lenkräder auch den Trend zu Elektroautos unterstützen, da sie im Stand beim Laden etwa mehr Platz im Innenraum böten.

Patrick Schäfer

Table of Contents

List of Authors

Tim Ahrenhold IAV GmbH, Gifhorn, Deutschland

Katsuhiko Arai Nissan Motor Co., Ltd., Atsugi City, Japan

Seda Aydogdu MdynamiX AG, Benningen, Deutschland

Stefan Battlogg Inventus Development GmbH, St. Anton, Österreich

Matthias Becker Munich University of Applied Sciences, München, Deutschland

Christoph Bittner Dr. Ing. h.c. F. Porsche AG, Weissach, Deutschland

Luka Brglez Elaphe propulsion technologies, Ljubljana, Slowenien

Paolo di Carlo Siemens Digital Industries Software, Leuven, Belgien

Torben Dittmar Institute for Automative Engineering (ika), RWTH Aachen, Aachen, Deutschland

Lutz Eckstein Institute for Automative Engineering (ika), RWTH Aachen, Aachen, Deutschland

Alexander Ein Waldt Ford Werke GmbH, Köln, Deutschland

Davide Ferrario Streparava Spa, Adro/Brescia, Italien

Taichi Fujita Nissan Motor Co., Ltd., Atsugi City, Japan

David Gagliardi Streparava Spa, Adro/Brescia, Italien

Gorazd Gotovac Elaphe propulsion technologies, Ljubljana, Slowenien

Alexander Haas Dr. Ing. h.c. F. Porsche AG, Weissach, Deutschland

Zoltan Hankovszki AVL List GmbH, Graz, Österreich

Johannes Hendewerk MdynamiX AG, München, Deutschland

Roman Henze Institute of Automotive Engineering (IAE), TU Braunschweig, Braunschweig, Deutschland

Matthias Holzleitner STIWA Group, Attnang-Puchheim, Österreich

Thomas Kersten Volkswagen do Brasil, São Bernardo do Campo, Brasilien

Tomaž Kompara Elaphe propulsion technologies, Ljubljana, Slowenien

Dennis Kurzweil Ford Werke GmbH, Köln, Deutschland

Hendryk Lausch Institute of Automotive Engineering (IAE), TU Braunschweig, Braunschweig, Deutschland

Michael Lipp AVL List GmbH, Graz, Österreich

Jan Loof AVL List GmbH, Graz, Österreich

Thomas Maier thyssenkrupp Presta AG, Eschen, Liechtenstein

Francesco Mega Streparava Spa, Adro/Brescia, Italien

Luca Mensi Streparava Spa, Adro/Brescia, Italien

Gregor Menze Dr. Ing. h.c. F. Porsche AG, Weissach, Deutschland

Simon Merz Schaeffler Technologies AG & Co. KG, Herzogenaurach, Deutschland

Gioele Micheli MydynamiX AG, Benningen, Deutschland

Jernej Munih Elaphe propulsion technologies, Ljubljana, Slowenien

Oliver Nehls Ford Werke GmbH, Köln, Deutschland

Matthias Niegl MdynamiX AG, München, Deutschland

Marco Occhineri Siemens Digital Industries Software, Leuven, Belgien

Peter Pfeffer MdynamiX AG, Benningen, Deutschland

Aleš Ručigaj Elaphe propulsion technologies, Ljubljana, Slowenien

Bernhard Schick Institut für Farerassistenz und automatisiertes Fahren, Benningen, Deutschland

Dieter Schramm Department Mechanical Engineering, Universität Duisburg-Essen, Duisburg, Deutschland

Kevin Schuler Institut für Farerassistenz und automatisiertes Fahren, Benningen, Deutschland

Michael Story Nexteer Automotive Corporation, Saginaw/MI, USA

Tobias Tarne MdynamiX AG, Benningen, Deutschland

Hiroyuki Tanaka Nissan Motor Co., Ltd., Atsugi City, Japan

Owen Tosh Nexteer Automotive Corporation, Saginaw/MI, USA

Pascal von Kamp Dr. Ing. h.c. F. Porsche AG, Weissach, Deutschland

Goetz-Philipp Wegner Ford Werke GmbH, Köln, Deutschland

Bernhard Wiesneth Schaeffler Technologies AG & Co. KG, Herzogenaurach, Deutschland

Using high-bandwidth direct-drive architecture for new motion control capabilities

Gorazd Gotovac[1], Jernej Munih[1] , Tomaž Kompara[1], Luka Brglez[1] and Aleš Ručigaj[1]

[1] Elaphe propulsion technologies, Teslova ulica 30, 1000 Ljubljana, Slovenia
gorazd.gotovac@elaphe-ev.com

Abstract. The capabilities of high-bandwidth direct-drive in-wheel motor (IWM) technology and its potential to revolutionize vehicle dynamics are explored. Elaphe's IWM technology offers unprecedented control over longitudinal, lateral, and vertical forces, presenting a significant breakthrough in vehicle motion control. This study demonstrates the advantages of IWM architecture, particularly in improving comfort and performance metrics through active body control. The paper details the mechanisms by which IWMs enhance vehicle dynamics, including faster torque delivery and improved traction control. Additionally, the impact of IWM on vertical dynamics is evaluated, showcasing its ability to mitigate the challenges related to increased unsprung mass. Through simulation and real-world testing, the research highlights the potential for IWMs to serve as integral chassis actuators, providing both performance and comfort improvements.

Keywords: in-wheel motors, vehicle motion control, reaction force control

1 Introduction

The advancement of modern automotive technologies requires innovative solutions to enhance vehicle performance, safety, and comfort. Elaphe's in-wheel motor (IWM) technology (Fig. **1**) presents an opportunity in the field of propulsion and motion control of vehicles. At the same time IWM is controversial in a large part due to the additional mass and related expected negative effects on ride dynamics. The aim of this paper is to demonstrate the former opportunity and show the path to mitigate the negative effects of the added mass. A key role in this research is played by suspension reaction forces, that occur due to the unique mounting of the in-wheel motor directly to the suspension members. These reaction forces are finely controllable at high bandwidth, which means they can be used in superposition to the low frequency traction and braking torque.
This paper aims to explore these direct-drive IWM architecture capabilities, on a simple case of a single bump. This case allows us to investigate gradually how the implementation of said principles improve the vehicle comfort, going from a passive system, to a controlled system and an optimized suspension and control.
In addition to compensating for negative effects we expect to generate new or significantly improved functionalities compared to conventional powertrain architectures. Other comfort and performance related proof of concept tests are shown as a potential for future work.

Fig. 1. Elaphe in-wheel motor layout

P. E. Pfeffer (Ed.), *15th International Munich Chassis Symposium 2024*,
Proceedings, https://doi.org/10.1007/978-3-662-71064-7_1

2 In-wheel motor motion control capabilities

The concept of in-wheel motors (IWM) has been extensively studied in recent years primarily as a means to improve powertrain packaging and vehicle overall energy consumption reduction. A significant part of investigations has also focused also on IWM as a possible way to optimize vehicle dynamics and improve overall performance. This is primarily based on the IWM ability to deliver torque much faster than a conventional e-drive, primarily due to the driveshaft properties [1]. When utilizing modern anti-jerk control a conventional e-axle can reach 90% of full torque in about 160 ms [2] while an in-wheel drive can increase torque at a rate of >500 Nm/ms (Fig. 2) and reach full torque in a factor of 20 lower time.

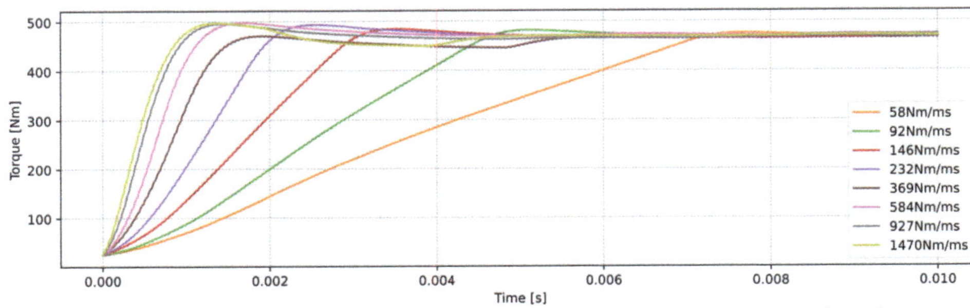

Fig. 2. Chart shows measured torque response for various control torque ramp rates (legend).

Investigations have shown that the superior bandwidth of direct drive machines allow for an improved stopping distance by as much as 15% in poor traction conditions, as well as improve the maximum vehicle sideslip angle in a double lane change scenario by as much as 20% [3]. At the same time, the bandwidth can be used to dampen longitudinal vibrations, resulting in a factor >3 improvement in RMS longitudinal jerk and an order of magnitude reduction of longitudinal acceleration performance indicators when road preview is available [4]. An important field of improvement is also traction control in poor traction conditions, as shown in Fig. 3.

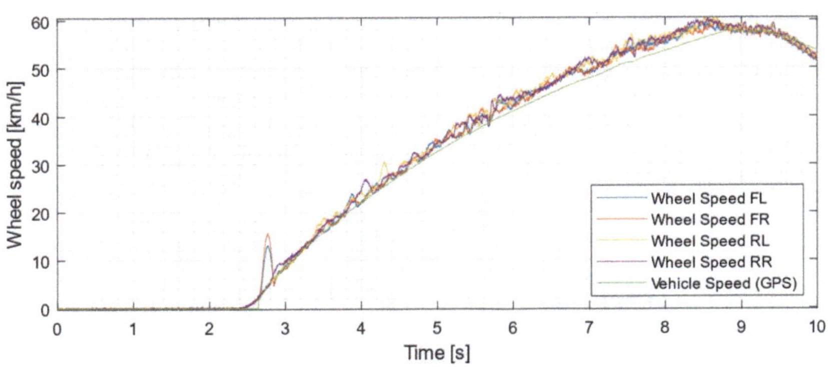

Fig. 3. Elaphe test vehicle traction control test on low-mu surface (wet painted surface).

In recent years research also evaluated the impact of IWM suspension reaction torque (Fig. 4) on chassis dynamics. Katsuyama et.al summarize in [5] the ability of a vehicle with 4×IWM to match the target performance of a vehicle with full active suspension and 4 wheel steering in scenarios such as step-steer, resulting in a 20% improvement in the planar motion natural frequency ω_n and the damping ratio ζ, a 50% improvement in roll and pitch damping coefficients and no change in steering characteristics. Other papers cover different aspects of the vertical dynamics controls for comfort or performance from the perspective of cornering comfort [7], ride comfort [8,9] and acceleration comfort [10] all showing the potential of in-wheel motors as chassis actuators.

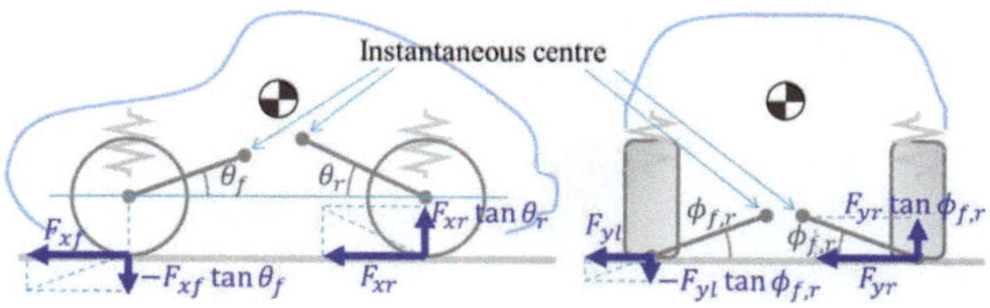

Fig. 4. Mechanism of vertical forces generated by tyre longitudinal and lateral forces [5]

In-wheel motors have traditionally been used in light, relatively low performance vehicles, but the vehicle dynamics advantages make them interesting also for high-performance vehicles. As the demand grows, the performance of in-wheel motors is also improving, today reaching up to 4500 Nm, 4000 rpm and 450 kW per wheel. Such performance opens up new discussion topics related to the ratio between the sprung and the unsprung weight.

3 IWM vehicle comfort and performance

In addition to the positive impact of the IWM propulsion architecture, one of the main arguments for the slow adoption of technology is the unsprung weight and the potential additional cost of managing it to keep the target performance and comfort. This paper sets out to first show using simulation the sensitivity of performance and comfort vs. added wheel weight; then demonstrate the same effect in a simplified scenario and finally use control methods to overcompensate effects of the unsprung mass both in simulation and in real vehicle measurements.

3.1 Unsprung weight sensitivity

A relatively light vehicle was used as a simulation base, weighing 1.64 tons. The sensitivity was performed by adding weight to the base vehicle unsprung weight in increments of 20 kg, at the same time keeping the total vehicle weight constant. At each step the suspension was retuned to match the original vehicle suspension natural frequency. Surface roughness ranged from smooth to Belgian brick road with maximal step between adjacent bricks of 14 mm.

Fig. 5. Slalom simulation in IPG Carmaker

The vehicle performance was evaluated based on two KPIs:

- Grip KPI (Vertical contact patch force RMS, see Fig. 6) and
- Comfort KPI (Combined RMS value of a_z, see Fig. 7)

Lower values are preferable for both KPIs, as a lower Grip KPI reduces vertical tire force oscillations and a lower Comfort KPI decreases oscillations of vertical acceleration.

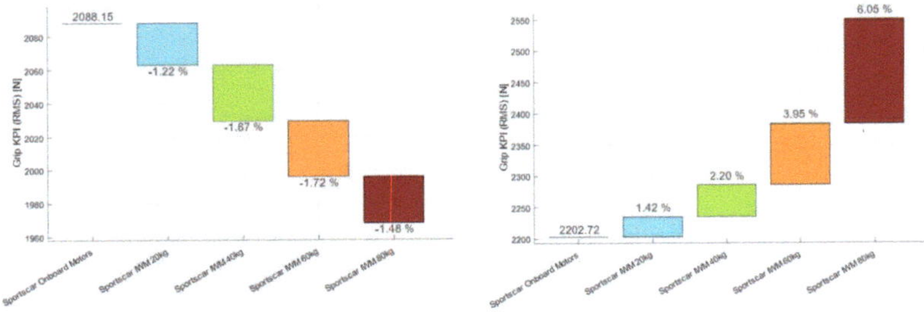

Fig. 6. Grip KPI evaluation for unsprung weight 20 kg increments (a) smooth road; (b) maximum simulated roughness. Lower values provide more consistent grip.

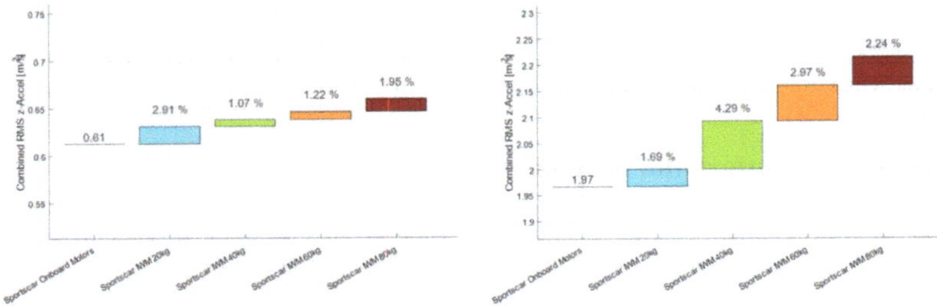

Fig. 7. Comfort KPI evaluation for unsprung weight 20 kg increments (a) smooth road; (b) maximum simulated roughness. Lower values provide better comfort.

The sensitivity analysis showed that the grip deteriorates, as surface roughness increases and that on smooth surfaces grip is better_with higher unsprung weight, due to reduced roll of the lighter body. At the same time the comfort changes on the rougher road can be perceived from simulations, while on a smooth road the difference in absolute terms is negligible. The sensitivity analysis concludes with an interesting question: whether the effects of unsprung weight can be compensated on rough surfaces.

3.2 Vertical control – comfort

In order to check the ability of an in-wheel motor system to compensate the unsprung weight and test it on the vehicle, one of Elaphe's mule vehicles was chosen. The vehicle is equipped with 4 in-wheel motors. The selected scenario was a speed-bump (60 cm length, 3 cm height) driven at 15 kph.

Fig. 8. Mule vehicle in real life and its digital twin in IPG carmaker

The following KPIs were followed during simulation and testing:

- RMS value of vertical acceleration of the chassis corners $a_{z,rms}$
- Max value of vertical acceleration of the chassis corners $a_{z,max}$
- Max value of vertical velocity of the chassis corners $v_{z,max}$

In simulations, all three KPIs were calculated for one front (FL) and one rear (RL) corner while during testing with the mule vehicle, accelerations were measured only at the rear coerners.

Simulations

The simulations were performed for the following setups:

- Vehicle with e-axle (base)
- Passive vehicle with IWMs
- Vehicle using IWM control (see Fig. 9)
- Vehicle using IWM control, with optimized front suspension (see Fig. 10)

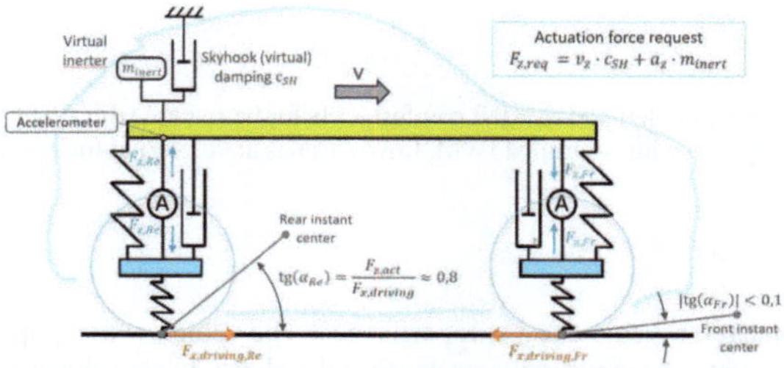

Fig. 9. Base vehicle schematic representation of active control.

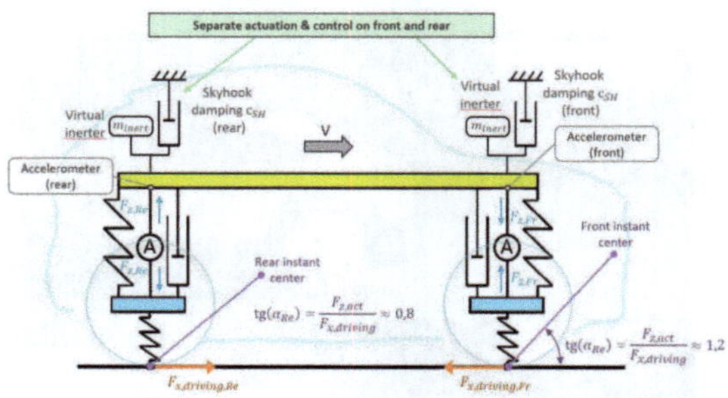

Fig. 10. Active control schematic for an optimized vehicle.

The addition of in-wheel motors (around +30 kg) has an expected small negative impact on the comfort KPIs (orange hue, Fig. 11).

Since the base vehicle has a virtually non-existent front reaction force (almost infinite front instant center), the controlled base vehicle with IWMs improves significantly only the rear behavior, while the front remains similar to the uncontrolled scenario (green hue, Fig. 11).

Finally, when the front suspension is modified to have a more favorable instant center, the behavior of the vehicle can be tuned between the front and the rear axis impact. One of possible tuning outcomes, focused more on the front passenger comfort is shown on the KPI chart (purple hue, Fig. 11). Potential non-favorable effects of such configuration were not analyzed.

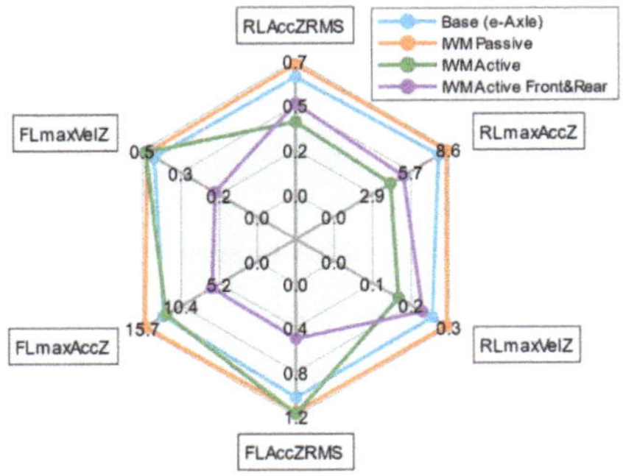

Fig. 11. Spider chart showing front left and rear left comfort KPIs for base, base with IWM, base with controlled IWM and optimized mechanical system with controlled IWM. Lower values are preferred for enhanced passenger comfort.

Vehicle measurements

To confirm the simulated behavior, vehicle test was performed. The vehicle was equipped with two accelerometers, one on each rear top-mount. The measurements confirm the ability to significantly impact the comfort KPIs (Fig. 14), by reducing peaks as well as the RMS value of acceleration and velocity in z direction (Fig. 13).

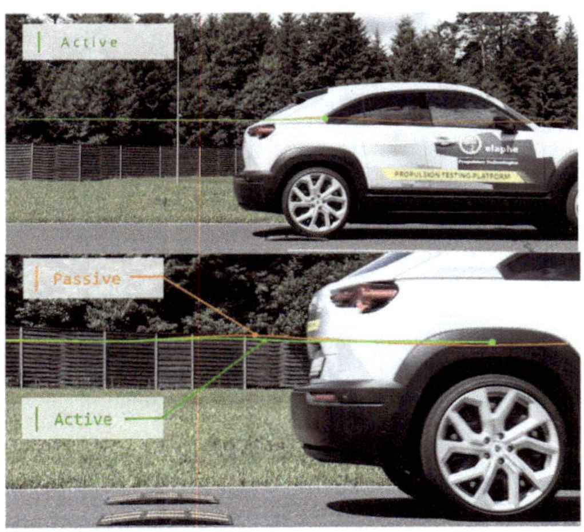

Fig. 12. Vehicle tests using Elaphe mule vehicle.

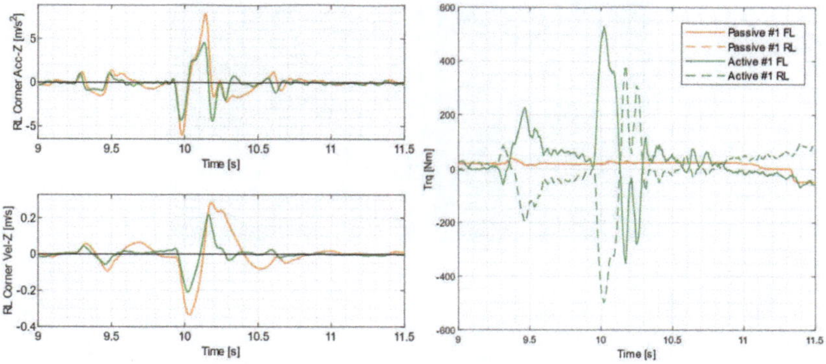

Fig. 13. Rear corner acceleration and velocity (left); front and rear torque profile (right).

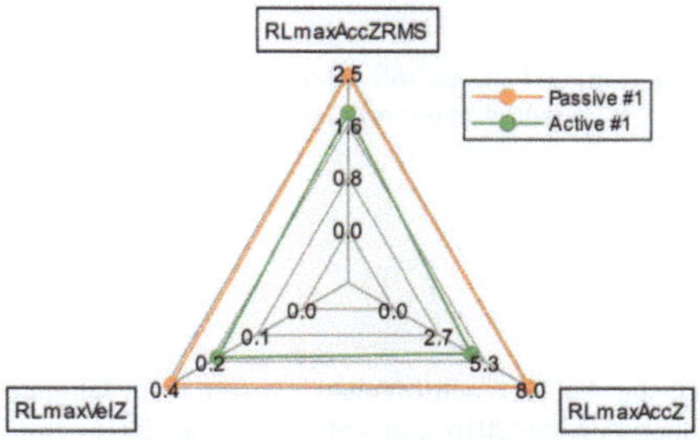

Fig. 14. Rear comfort KPIs improvement with active control of IWM motors. Lower values are preferred for enhanced passenger comfort.

4 Discussion and future work

Simulation and testing in this paper have demonstrated the potential of in-wheel motors to improve vehicle ride and comfort and counter the effects of the additional unsprung mass. Virtual experiment with a modified suspension showed the path towards an even more capable control. However as shown in the unsprung mass sensitivity, comfort is only a part of the story, so future work on simulation and testing of control of the vertical contact patch force variation is needed. First simulations in simple scenarios are promising (Fig. 11) but require refinement and physical validation.

Finally testing in complex scenarios will be required, to understand behavior of longitudinal, lateral, and vertical dynamics with IWM control. A slalom scenario plays an important role in this investigation, as going from a test on a smooth surface (Fig. 15) to a test on different surface roughness will require controlling roll as well as pitch at the same time.

The full potential of in-wheel motors for performance applications will however be achieved, when combined with other chassis actuators such as active damping, active camber and independent wheel steering.

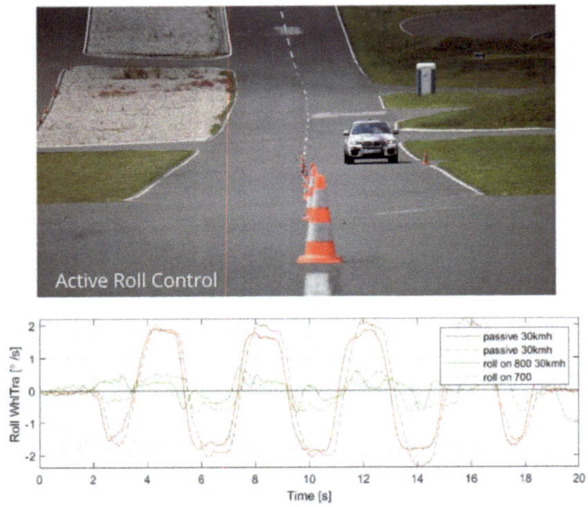

Fig. 15. Mule vehicle on test track (top) and chassis roll values (bottom) during slalom scenario on a smooth surface, with and without roll control with IWM.

References

1. V. Hermansson, E. Moparthi: Control of an Electric Vehicle Powertrain to Mitigate Shunt and Shuffle, Chalmers University of Technology, Master's thesis EX056/2016, Gothenburg, Sweden 2016
2. Liu, H., Chen, Y., Zhang, X., Li, W.: Shaft Torque Compensation for Electric Vehicle Driveline Oscillation Active Damping with Feedforward and PID Feedback Controller. In: Tan, J., Gao, F., Xiang, C. (eds) Advances in Mechanical Design. ICMD 2017. Mechanisms and Machine Science, vol 55. Springer, Singapore
3. Bayar, K.: Performance comparison of electric-vehicle drivetrain architectures from a vehicle dynamics perspective. Proceedings of the Institution of Mechanical Engineers, Part D: Journal of Automobile Engineering, 1-21 (2019)
4. V. Vidal et al.: On Pre-Emptive In-Wheel Motor Control for Reducing the Longitudinal Acceleration Oscillations Caused by Road Irregularities, in IEEE Transactions on Vehicular Technology, vol. 71, no. 9, pp. 9322-9337 (Sept. 2022)
5. Etsuo Katsuyama, Makoto Yamakado & Masato Abe: A state-of-the-art review: toward a novel vehicle dynamics control concept taking the driveline of electric vehicles into account as promising control actuators, Vehicle System Dynamics, 59:7, 976-1025 (2021)
6. Katsuyama E. Decoupled 3D moment control using in-wheel motors. Veh Syst Dyn J. 2013;51(1):18–31.
7. N. Ochi, H. Fujimoto and Y. Hori: Proposal of roll angle control method using positive and negative anti-dive force for electric vehicle with four in-wheel motors, 2013 IEEE International Conference on Mechatronics (ICM), Vicenza, Italy, pp. 816-821 (2013)
8. Katsuyama, E. and Omae, A.: Improvement of Ride Comfort by Unsprung Negative Skyhook Damper Control Using In-Wheel Motors, SAE Int. J. Alt. Power. 5(1):214-221 (2016)
9. Katsuyama, E: Improvement of ride comfort by triple Skyhook control. In: Pfeffer, P. (eds) 9th International Munich Chassis Symposium 2018. Proceedings. Springer Vieweg, Wiesbaden.
10. Kopylov, Semen & Phanomchoeng, Gridsada & Ambrož, Miha & Petan, Žiga & Kunc, Robert & Qiu, Yi.: Improvements to a vehicle's ride comfort by controlling the vertical component of the driving force based on in-wheel motors. Journal of Vibration and Control. 29 (2022).

OBWE - On Board Weighing Equipment

Luca Mensi[1], Francesco Mega[1], Davide Ferrario[1], David Gagliardi[1]

[1] Streparava Spa, Via Zocco 13, 25030 Adro (Brescia), Italy
l.mensi@streparava.com
f.mega@streparava.com
d.ferrario@streparava.com
david.gagliardi_consultant@streparava.com

Abstract. On Board Weighing systems serve as a preventative measure against vehicle overloading, enhancing both safety and environmental conservation by diminishing pollution and energy usage. This contributes to the preservation of road infrastructure and the promotion of equitable competition. Streparava has developed a system that employs a hybrid methodology to swiftly and accurately assess the load and its distribution within a vehicle. This system ensures precise and dependable measurements, which also facilitates the incorporation of extra functionalities for the vehicle.

Keywords: OBW, LCV, vehicle weight assessment

1 Introduction

The trucking industry faces a critical issue with vehicle overloading. A substantial portion of violations pertains to vehicle weights, with statistics showing that one out of every three inspected vehicles is carrying more than the legal limit. Often, these vehicles are laden with 10 to 20% more weight than permitted. Such overloading has detrimental effects on the environment, as it leads to increased pollution and energy usage proportional to the vehicle's weight. Moreover, overweight vehicles are the root cause of various adverse outcomes, including compromised road and driver safety, accelerated road wear, environmental damage, and disruption of fair market competition.

In response to these challenges, governments are actively developing regulations to curb these violations. The European Union, for instance, has already put forth a legislative proposal [1].

Stakeholders in the trucking sector have diverse viewpoints and requirements regarding weight compliance.

Public authorities aim to enhance their overarching objectives, which include minimizing road maintenance costs, reducing the environmental footprint of road transport, bolstering road safety, and optimizing the road transport system's efficiency.

Transport companies seek to transport goods effectively without the hindrance of excessive administrative procedures, paperwork, or frequent stops for weight checks. They also advocate for equitable competition, emphasizing the consistent enforcement and prosecution of regulations across all market participants.

P. E. Pfeffer (Ed.), *15th International Munich Chassis Symposium 2024*,
Proceedings, https://doi.org/10.1007/978-3-662-71064-7_2

Vehicle manufacturers strive to generate profit by offering their clientele innovative, productivity-boosting, and unique products.

Lastly, professional drivers aspire to perform their duties in conditions that ensure safety and comfort, both within their vehicle and on the roadways.

To mitigate the issue of vehicle overloading, two distinct technological approaches are available: Weigh-in-Motion (WIM) and On-Board Weighing (OBW). WIM systems necessitate integration with road infrastructure and, depending on the desired level of precision, can be expensive to implement. They also require regular maintenance. To achieve a high density of compliance checks, it's essential to install a substantial number of WIM systems. In contrast, OBW systems are installed directly into vehicles. This allows for the transmission of weight data to various stakeholders, such as authorities conducting roadside inspections, the vehicle's driver, or fleet management systems, at any time, even while the vehicle is in transit.

2 Streparava OBWE

Streparava has introduced a cutting-edge OBW (On-Board Weighing) system capable of accurately determining the total weight of a vehicle (GVW - Gross Vehicle Weight) and the distribution of this weight across each axle and wheel [2]. Utilizing a sophisticated algorithm, the Streparava OBWE employs a data fusion method to combine and analyze data from the vehicle's CAN bus, sensor inputs, and driving conditions to deliver precise load information. This system is primarily designed for light commercial vehicles (LCV), yet its flexible and adaptable nature makes it applicable to a variety of vehicles, including heavy trucks. Additionally, the system is equipped to assess the Center of Gravity (CoG) position. Targeted primarily at vehicle manufacturers, the Streparava OBWE also offers seamless aftermarket integration for existing vehicles.

2.1 Layout Configuration

Fig.2 and Fig. 2 illustrate the components of the OBWE system, which includes:

- Sensors for wheel travel and/or air spring pressure at each corner of the vehicle.
- An Electronic Control Unit (ECU) that processes data from various sources.
- An Inertial Measurement Unit (IMU) affixed to the vehicle's chassis and embedded into the ECU.
- An estimation algorithm programmed into the ECU.

The function of the wheel travel sensors is to ascertain the relative positioning of the wheel center to the chassis at each corner, while the pressure sensors keep track of the air springs, particularly in vehicles with pneumatic suspension systems.

The IMU plays a crucial role in gauging the vehicle's dynamic state, including roll, pitch, yaw angles, and chassis accelerations, necessitating its installation on the vehicle chassis and integration with the ECU.

The ECU gathers information from the sensors and the CAN bus data, such as vehicle speed, gear position, and brake engagement. It then employs a data fusion method to deduce and relay mass-related data, typically back to the vehicle's CAN bus.

2.2 Algorithm Integration

Fig. 4 illustrates the OBWE algorithm's structure, which is divided into several components:

- Data Handling: This module takes charge of filtering and aligning data from sensors and the vehicle's CAN bus to prepare for subsequent processing stages.
- Movement Analysis: By examining the data from the Data Handling module, this system component can determine the vehicle's current state and decide whether to proceed with further assessments. It selects maneuvers that ensure accurate weight readings, focusing on data from specific driving conditions such as non-braking, minimal acceleration, and gentle slopes, among others.
- Suspension Parameters: This part of the system uses Look-Up Tables (LUTs) to define critical suspension-related variables. It calculates the vehicle's mass and its distribution by correlating the wheel's vertical movement and the load on the ground, using the LUTs embedded in the software. The results are then forwarded to the subsequent module.
- Data Consolidation: Prior to finalizing the output, this module aggregates and processes the data. It uses a group of data points rather than individual ones to determine the final output.

The vehicle's suspension system is represented within the ECU using specific LUTs, which, after an initial characterization of the suspension system and considering overall tolerances and data distribution, require calibration through an End of Line (EOL) procedure.

Looking ahead, Streparava OBWE aims to incorporate an automatic calibration process to streamline product integration.

An initial assessment of the Gross Vehicle Weight (GVW) can be made even when the vehicle is not in motion, but the accuracy of this measurement is somewhat compromised by factors such as friction, hysteresis, and deformations affecting the vehicle's components.

While the vehicle is in motion, the integrated algorithm within the Engine Control Unit (ECU) monitors the car's dynamic condition, discerns the driver's intended maneuvers, isolates the aforementioned occurrences, and swiftly adjusts the mass-related metrics, typically in under ten seconds.

2.3 Test and validation

To validate the system's performance and load estimation accuracy, extensive testing was conducted using various suspension setups and load conditions. The tests demonstrated the system's high precision, as evidenced by Fig. 3 and Fig. 5, with an accuracy surpassing ±2.5%, which exceeds the EU 2019/1213 standard of ±5% at 90% GVW. The Streparava OBWE boasts a 10kg resolution, a 10ms sampling rate, and delivers initial estimates within 10 seconds.

The system consistently estimates the load at each corner of the vehicle with the previously mentioned accuracy. By accurately modeling suspension components such as springs, bump stops, dampers, and anti-roll bars, it can precisely calculate the vehicle's load and its distribution, regardless of how the load is spread, the vehicle's maneuvers, or driving style.

The OBWE is adept at recognizing specific driving maneuvers and compensating for load imbalances caused by road inclines and chassis movement during travel. It can also detect shifts in load positioning while the vehicle is in motion. When compared to alternative methods, the Streparava OBWE offers seamless integration with any suspension system.

2.4 Features enabled

Various OBWE systems are available in the market, each employing distinct methodologies. Some utilize advanced vehicle modeling techniques; however, these often lack reliability due to the variability in physical and vehicular conditions. Others rely on technologies like strain gauges, necessitating extensive suspension characterization for accurate results. The Streparava OBWE system adopts a hybrid strategy. It employs a vehicle dynamics model to determine the specific conditions under which measurement data is considered dependable, informed by an initial suspension analysis at the end of the production line (EOL).

Rather than attempting to simulate every possible vehicle maneuver to continuously estimate the vehicle's weight, the Streparava approach involves periodic verification, activated only when the model's predictions are deemed trustworthy. The Streparava OBWE system communicates load information via the CAN bus, and also provides data on load distribution and the center of gravity's coordinates (x,y,z). This information can enhance other vehicle systems by enabling features such as:

- Improved Tire Pressure Monitoring: Tire pressure recommendations are adjusted based on the individual tire's load.
- Adaptive Retarder: Utilizes mass and distribution data to automatically set the retarder torque.
- Unbalanced Load Alerts: Monitors the distribution of mass between the left and right sides to prevent vehicle drift and uneven wear on brakes and tires.
- Adaptive Electronic Brake-force Distribution (EBD): Employs mass distribution data to fine-tune brake torque allocation, enhancing the rear axle's effectiveness.
- Overturning Risk Notifications: Uses the calculated center of gravity height to alert drivers of potential overturn risks.
- Fleet Management: Enables optimization of fleet cargo capacity.
- Load and Mission Profile Database: Allows Original Equipment Manufacturers (OEMs) to create a database of mission profiles based on actual load data usage.

2.5 Future Improvements

To enhance the system's capabilities and oversee its performance, integration of a watchdog algorithm is planned. The system will utilize data from the vehicle's Controller Area Network (CAN), applying a machine learning methodology to deduce the stiffness of the suspension. This process of estimation takes considerably longer (several minutes) compared to the brief duration required for load estimation (a few seconds). However, throughout the operational period of the vehicle, this stiffness estimation can serve to corroborate and refresh the lookup table data embedded in the software. The watchdog feature will facilitate ongoing surveillance of the On-Board Weighing Equipment (OBWE) system's condition, pinpointing any deviations or manipulations of the suspension.

While this method is not a substitute for the primary algorithm due to its lesser precision and slower response time, it proves beneficial in detecting inconsistencies that may arise from wear and tear of the suspension system. By harnessing the comprehensive data gathered through an AI-driven approach, it is feasible to activate diverse features and incorporate an automatic calibration process.

3 Summary and outlook

The OBW system is designed to provide critical alerts for axle and gross weight overloads, ensure optimal load distribution between the front and rear, enhance the vehicle's payload capacity, and diminish wear and tear as well as maintenance needs. The Streparava OBWE offers a range of benefits, including environmental advantages by reducing pollutant emissions through the prevention of overloading and improved load distribution. It enhances safety by mitigating the risks associated with overloading and potential vehicle rollovers. The system boasts seamless vehicle integration, adaptability to various suspension setups, and delivers precise and dependable weight estimations across the entire load spectrum. Additionally, it can calculate the vehicle's center of gravity (CoG), providing valuable data that could support extra functionalities. The OBWE's watchdog feature is adept at identifying any tampering or irregularities within the suspension system. Lastly, it contributes to maintenance by utilizing load data to inform a precise predictive maintenance strategy for the vehicle.

Figures

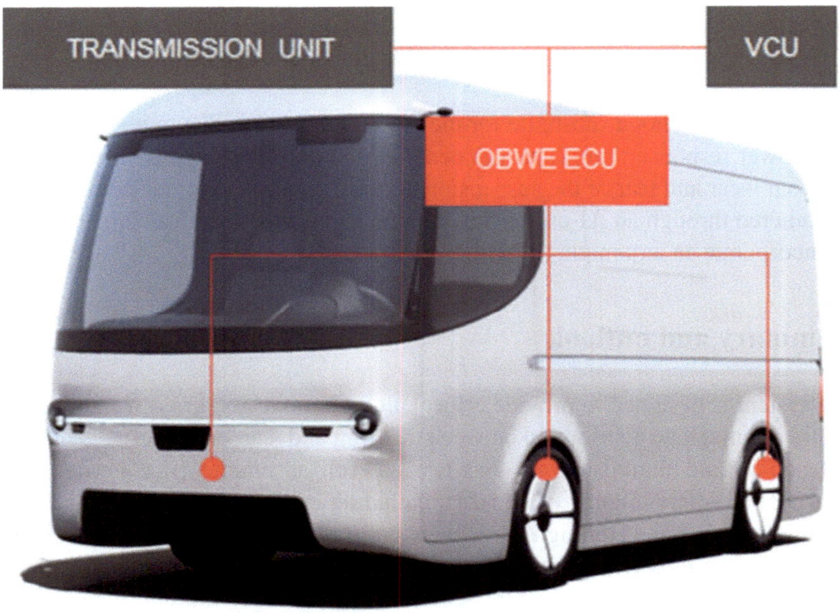

Fig. 1. Streparava OBWE block scheme connection (© Free art Director, Getty Images/iStock | Streparava Spa)

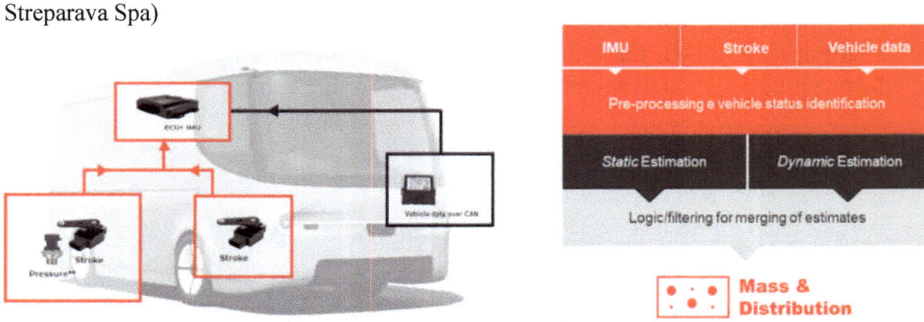

Fig.2. System layout, on the left in red the hardware elements that compose OBWE. On the right is reported the estimation process flow. (© Free art Director, Getty Images/iStock | Streparava Spa)

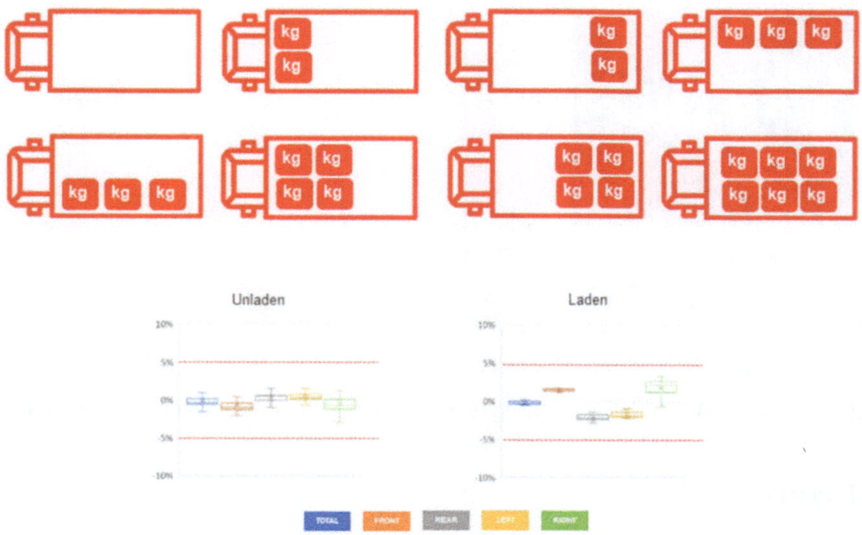

Fig.3. On the top a set of test load configurations. On the bottom is reported the accuracy of load estimation for a two axles LCV (laden and unladen). GVW in blue, front axle in orange, rear axle in grey, yellow for left side and green for right side. (© Streparava Spa)

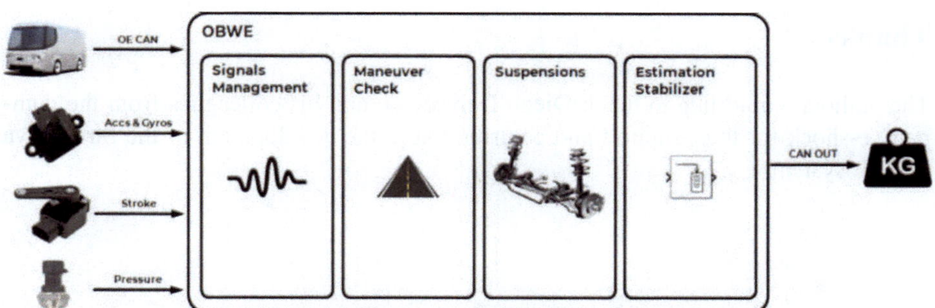

Fig.4. Algorithm structure with the different modules (© Free art Director, Getty Images/iStock | Streparava Spa)

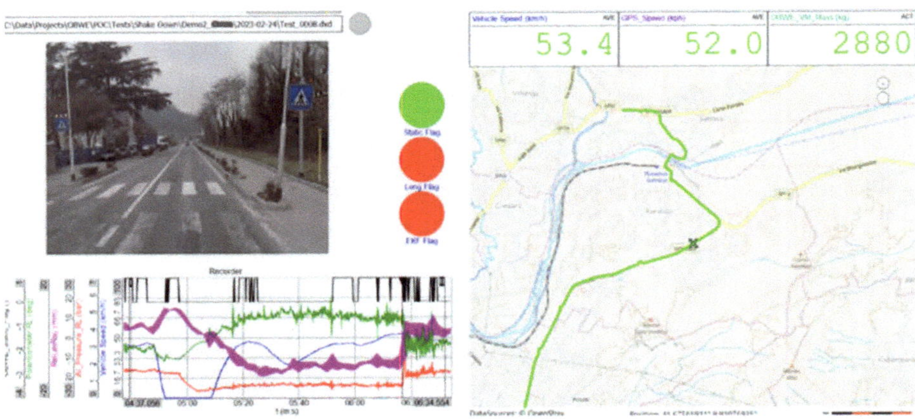

Fig.5. Example of OBWE interface test system used during the validation (© Streparava Spa)

References

[1] EU Regulation 2019/1213, 12 July 2019. http://data.europa.eu/eli/reg_impl/2019/1213/oj

[2] PCT Application No. PCT/IB2023/058225, Method for calculating the mass of a vehicle and system for calculating the mass of said vehicle, Aug.2023

Thanks

The authors would like to thank Diego Delvecchio and his colleagues from the company e-shock for their support and contribution to the development of the Streparava OBWE system.

Potential analysis of semi-active anti-roll bar based on MR technology

Tobias Tarne[1], Peter Pfeffer[1] and Stefan Battlogg[2]

[1] MdynamiX AG, Junkersstr. 4, 87734 Benningen, Germany
[2] Inventus Development GmbH, Montafonerstr. 68, 6771 St.Anton, Austria

Abstract. This paper explores the potential of semi-active anti-roll bars utilizing magnetorheological (MR) technology. It begins with a detailed discussion of MR technology's principles, followed by an examination of its integration into semi-active anti-roll bar systems. The experimental setup includes dynamic driving tests comparing the performance of semi-active anti-roll bars against passive and fully active systems. MR technology enables rapid adaptation of the anti-roll bar's stiffness and damping properties to match driving conditions, thereby enhancing vehicle stability and ride comfort during dynamic maneuvers.

Performance evaluation is based on vehicle dynamics tests conducted on a proving ground, assessing the semi-active system's influence on vehicle handling characteristics such as roll angle and understeer behavior. Results demonstrate that the semi-active system, despite its simplicity and lower power requirements compared to fully active systems, provides significant improvements in handling dynamics. These benefits are achieved without the complexities and higher costs associated with fully active systems.

In conclusion, the paper highlights the advantages of semi-active anti-roll bars in improving driving dynamics and comfort. The technology offers a promising alternative for future automotive applications, providing a balanced approach between performance enhancement and resource efficiency. The study underscores the semi-active system's potential as a viable solution for enhancing vehicle dynamics, especially in mid-range and economy vehicles.

Keywords: Semi-active Anti-Roll Bar, Magnetorheological Technology, Vehicle Dynamics, Ride Comfort, Stability Control.

P. E. Pfeffer (Ed.), *15th International Munich Chassis Symposium 2024*,
Proceedings, https://doi.org/10.1007/978-3-662-71064-7_3

1 Motivation

To enhance the comfort of passengers, the vehicle's suspension should be sufficiently soft to absorb road irregularities. To counteract the high roll angle resulting from the soft suspension, stabilizers also known as anti-roll bars are fitted. The anti-roll bar limits the vehicle's roll angle and provides the ability to adjust its understeering gradient. With installed anti-roll bar, the wheels are no longer independent. In straight line driving the anti-roll bar has hardly any influence if the left and right-side wheel are displaced by the same amount and in the same direction. However, if the vehicle encounters a speed bump with only one wheel, a force is also applied to the other wheel, leading to more body movement than desired.

Almost all vehicles are equipped with passive anti-roll bars, which lack adjustability without replacing mechanical components. As a result, the suspension system has characteristics that are not adaptable to varying scenarios. When considering the application of a passive suspension system, it's important to carefully consider the trade-offs, despite its cost-effectiveness. While passive suspension systems are economical, they necessitate compromises in their implementation.

The use of semi-active or active anti-roll bars offers advantages, even when integration into a production vehicle is not initially planned. During the prototype vehicle testing and development phase, these systems facilitate a more rapid determination of the optimal stiffness for the anti-roll bar, enhancing the tuning process. By employing the more cost-effective and compact semi-active system, testing and development drivers can assess various stiffness settings of the anti-roll bar directly. This approach not only achieves cost savings but also improves the tuning results. Additionally, it eliminates the necessity to produce multiple anti-roll bars with different stiffness values, streamlining the development process.

To evaluate the potential of a semi-active anti-roll bar system, vehicle dynamics tests are conducted on a proving ground using a vehicle equipped with conventional anti-roll bars. Following these initial tests, the vehicle's passive anti-roll bars are replaced with the Inventus semi-active anti-roll bars, and identical tests are carried out. The semi-active roll bars are engineered to match the torsional stiffness of the original passive system. Although the primary objective was not to reduce roll angles but rather to enhance comfort, the analysis concludes with a comparative assessment of the semi-active anti-roll bars against a fully active suspension system.

2 Passive, semi-active and active anti-roll bar

Active suspension systems offer several benefits over the passive suspension counterparts, including the ability to adapt to varying road conditions and driving scenarios. One key advantage is the system's ability to adjust the suspension of the vehicle within milliseconds to prevailing road conditions. The vehicle's roll angle can be dynamically adjusted or reduced, and in some cases even inverted, as seen in the Audi S8's 'comfort+' mode, which can increase passenger comfort [1]. Furthermore, the vehicle's understeering and oversteering behavior can be dynamically adjusted using hydraulic or electromechanical motors in the active anti-roll bar (Fig. 1) that create forces on each wheel via the anti-roll bar. By virtually adjusting the torque of the anti-roll bar in different situations, the decision-making in scope of trade-offs can be minimized. This reduces the need to compromise between driving comfort on straight roads and vehicle dynamics during cornering. As a result, active suspension systems and active anti-roll bars offer a more adaptable driving experience.

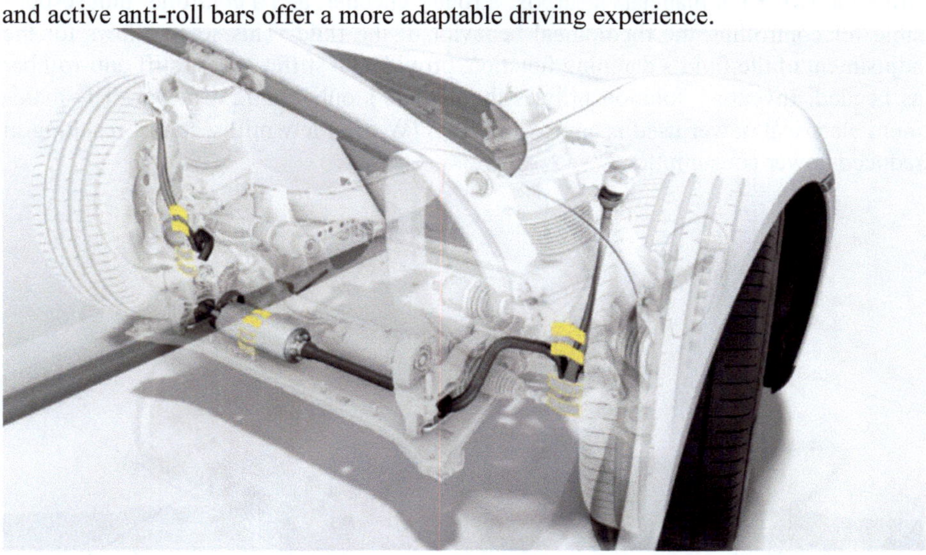

Fig. 1. Active anti-roll bar system in the Porsche Taycan [2].

Semi-active anti-roll bars mainly consist of an enhanced passive system featuring an adjustable torque element. Semi-active anti-roll bars do have advantages over both active and passive systems. Their simpler design requires less power and design space compared to active systems, making them a practical and cost-effective choice for a broad range of vehicles. This includes mid-range and economy models as well as vehicles without sufficient design space for active systems. It is important to note that the semi-active anti-roll bars can only adjust the stiffness and damping of the anti-roll bar and cannot generate torque without a twist angle of the anti-roll bar, unlike the active system mentioned earlier. Therefore, it is not possible to cancel or to invert the roll angle, as it is done in active anti-roll bar systems.

Inventus has developed a semi-active anti-roll bar based on Magnetorheological Fluid (MRF) technology to adjust the stiffness and damping of the anti-roll bar in milliseconds. The semi-active anti-roll bar's structure comprises an actuator unit that contains the MRF, a planetary gearbox, a rotary encoder, and a controller unit responsible for controlling the rheological behavior of the fluid. The MRF is a mixture of a base oil and magnetizing iron particles, typically ranging in size from 1-10 micrometers. When a magnetic field is applied to the fluid, the iron particles align themselves along the magnetic flux within milliseconds, forming particle chains. The concentration of particles on the magnetic lines and the resulting chain strength increases with the applied current and magnetic field strength, leading to a higher anti-roll bar stiffness. In the event of an electronic failure, the anti-roll bar will continue to function as a conventional anti-roll bar with normal function and its pre-designed stiffness, resulting in a fail-safe system. The semi-active anti-roll bar structure consists of a stiffer conventional anti-roll bar which is cut in the center, an actuator unit containing the MR fluid (MRF), a planetary gearbox, a rotary encoder, and a controller unit responsible for controlling the rheological behavior of the fluid. This setup allows for the adjustment of the fluid's damping function, providing a stiffer or less stiff anti-roll bar as needed. Inventus' solution utilizes the vehicle's onboard 12V power. The maximum electrical power used is approximately 37W for a few milliseconds, resulting in reduced power consumption.

Fig. 2. Semi-active anti-roll bar designed by Inventus.

3 Vehicle tests

The vehicle tests were conducted on a proving grounds skid pad with two different vehicles. An SUV with active suspension and a BMW X6 with its production anti-roll bars, which were later replaced by the MR anti-roll bars designed by Inventus. The torsional stiffness of the torsion bar in the semi-active anti-roll bar closely matched that of the production version. The BMW X6 F16 xDrive30d is a 2017 model and falls under the category of Sport Utility Vehicle (SUV). To minimize external influences during the tests, the Electronic Stability Control (ESC) system is deactivated. The ESC system can be an issue during testing due to the changing vehicle behavior as the stiffness of the front and rear anti-roll bars changes. For instance, if the car tends to oversteer more than with the passive anti-roll bars, the ESC system will likely attempt to stabilize the car. To fully evaluate the semi-active system's potential, the stiffness and damping was set to either the maximum or minimum possible.

To be able to do an objective comparison between the systems, the vehicles were equipped with various sensors. To measure speed and side slip angle the S-Motion Correvit from Kistler is used, while an inertial measurement unit (ADMA G-Pro+ from GeneSys) is mounted close to the vehicles center of gravity to record the vehicle body's accelerations. The vehicles center of gravity is measured and implemented as a point-of-interest in the ADMA system. This enables the calculation of the accelerations at this point. The inertial measurement unit provides the absolute roll and pitch angles, to compute the relative roll and pitch angles regarding the road surface, three laser sensors are mounted. The vehicle's original steering wheel is replaced by an SR30 steering robot from ABD. This provides repeatable and precise inputs that a human is not able to do with this accuracy.

4 Vehicle Test Results

To evaluate the performance and the potential of the semi-active system, the evaluation and benchmark software, MXeval is used. This software facilitates fully automated test evaluation and comparison of different systems based on Key Performance Indicators (KPIs). The use of MXeval minimizes analysis time and costs, as tests are automatically analyzed upon completion. Test results can be viewed within seconds, allowing for an immediate determination of whether the test run was successful or if repetition is necessary.

One of the primary objectives of using active or semi-active systems is to control the vehicle's roll angle. The roll angle refers to the vehicle's rotational motion around its longitudinal axis, which is caused by lateral acceleration in conjunction with a lever-arm to the vehicle's roll center.

To evaluate the changes in the roll angle, the slowly increasing steer as well as the weave test maneuver were conducted. During the slowly increasing steer the vehicle is driven at constant speed and a slow, steady steering wheel velocity is applied until reaching the vehicle's maximum lateral acceleration [3]. For the weave tests the vehicle is driven at constant speed while a sinusoidal steering input is applied, according to the required steering wheel angle, to produce a lateral acceleration of 2 m/s^2 and 4 m/s^2. The frequency of the sine is 0.2 Hz [4].

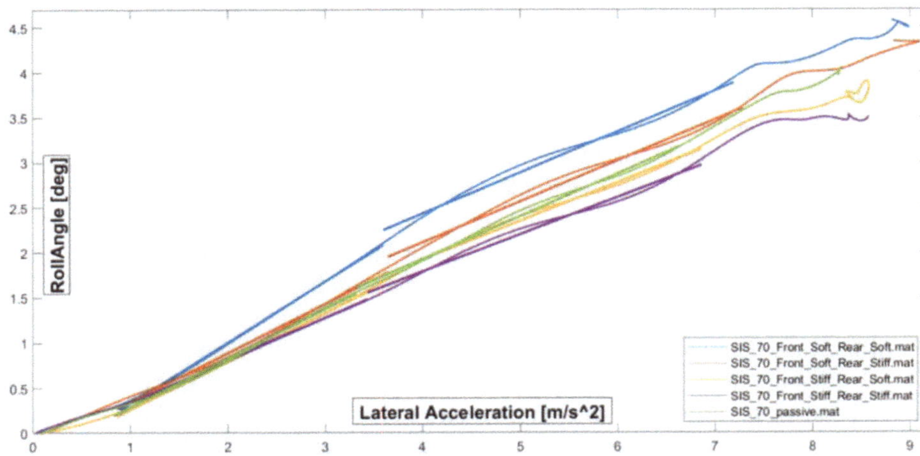

Fig. 4.1. Slowly increasing steer maneuver at 70 kph.

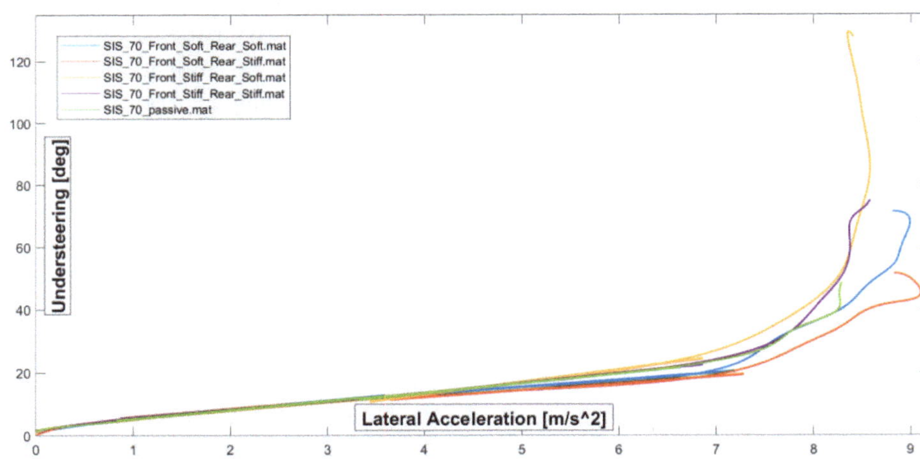

Fig. 4.2. Understeering during the slowly increasing steer maneuver at 70 kph.

Table 4.1. Slowly increasing steer KPIs at 70 kph.

KPI	Unit	Reference (passive system)	Front Soft Rear Soft	Front Soft Rear Stiff	Front Stiff Rear Soft	Front Stiff Rear Stiff
Maximum RollAngle	deg	4.03	4.72 (+17.23 %)	4.59 (+13.98 %)	3.98 (-1.17 %)	3.57 (-11.27 %)
RollAngle Gradient lowG	$\frac{deg}{\left(\frac{m}{s^2}\right)}$	0.55	0.68 (+23.64 %)	0.54 (-1.82 %)	0.53 (-3.64 %)	0.47 (-14.55 %)
RollAngle gradient highG	$\frac{deg}{\left(\frac{m}{s^2}\right)}$	0.47	0.45 (-4.26 %)	0.45 (-4.26 %)	0.43 (-8.51 %)	0.41 (-12.77 %)
Understeer gradient lowG	$\frac{deg}{\left(\frac{m}{s^2}\right)}$	2.80	2.82 (+0.71 %)	2.44 (-12.86 %)	2.71 (-3.21 %)	2.51 (-10.36 %)
Understeer gradient highG	$\frac{deg}{\left(\frac{m}{s^2}\right)}$	3.21	2.28 (-28.67 %)	2.19 (-31.78 %)	3.99 (+24.30 %)	3.13 (-2.49 %)

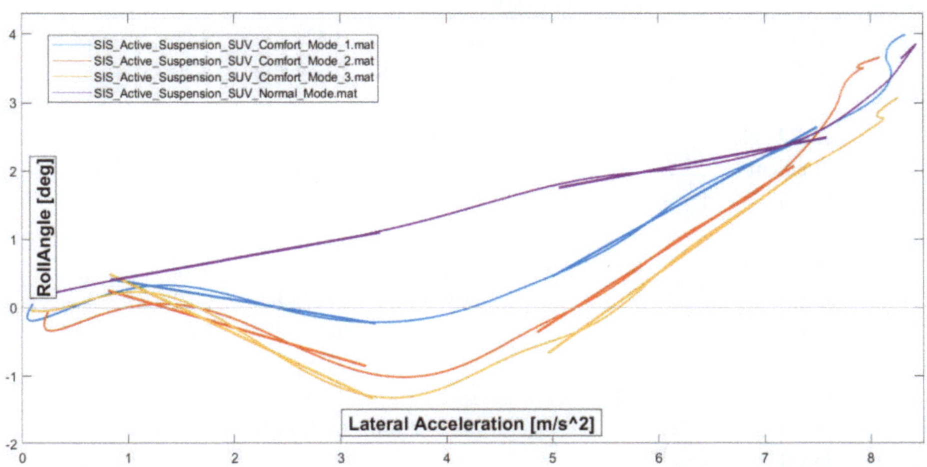

Fig. 4.3. RollAngle behavior in the slowly increasing steer maneuver at 70 kph with an active suspension vehicle.

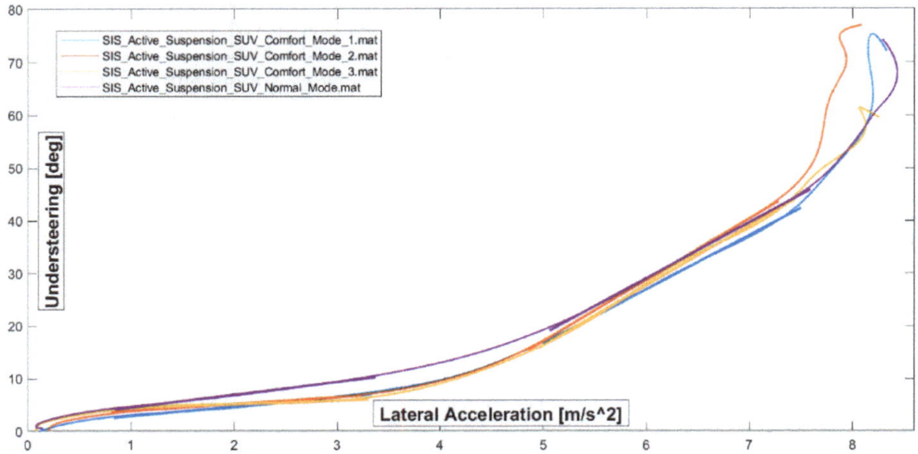

Fig 4.4. Understeering during the slowly increasing steer maneuver at 70 kph with an active suspension vehicle.

Table 4.2. Slowly increasing steer KPIs for an active suspension vehicle.

KPI	Unit	Reference (normal mode)	Mode 1	Mode 2	Mode 3
Maximum RollAngle	*deg*	3.87	4.10 (+5.94 %)	3.62 (-6.46 %)	3.06 (-20.93 %)
RollAngle Gradient lowG	$\frac{deg}{\left(\frac{m}{s^2}\right)}$	0.28	-0.26 (-192.86 %)	-0.46 (-264.29 %)	-0.74 (-364.29 %)
RollAngle gradient highG	$\frac{deg}{\left(\frac{m}{s^2}\right)}$	0.29	0.87 (+200 %)	1.00 (+244.83 %)	1.12 (+286.21 %)
Understeer gradient lowG	$\frac{deg}{\left(\frac{m}{s^2}\right)}$	2.15	1.51 (-29.77 %)	1.04 (-51.63 %)	0.40 (-81.40%)
Understeer gradient highG	$\frac{deg}{\left(\frac{m}{s^2}\right)}$	7.93	8.29 (+4.54 %)	8.95 (+12.86 %)	9.13 (+15.13 %)

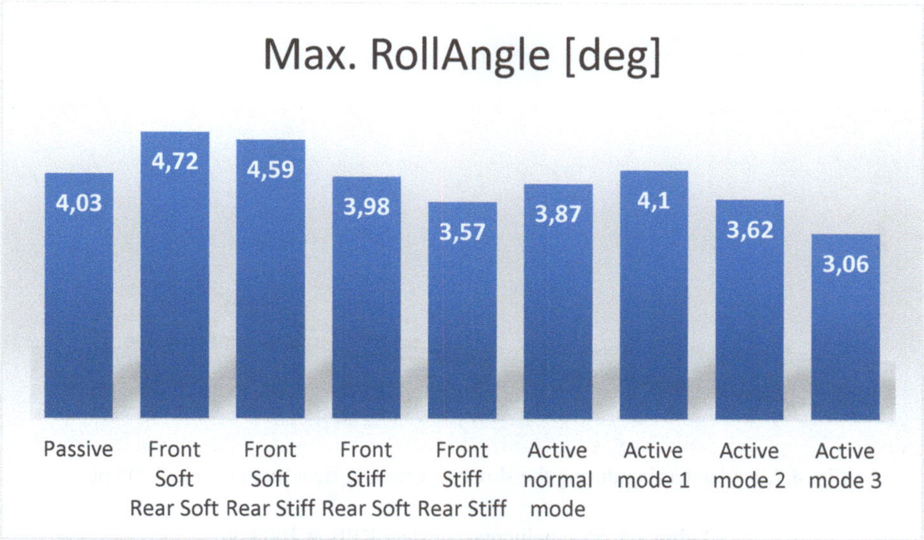

Fig. 4.5. Maximum RollAngle KPI in slowly increasing steer at 70 kph.

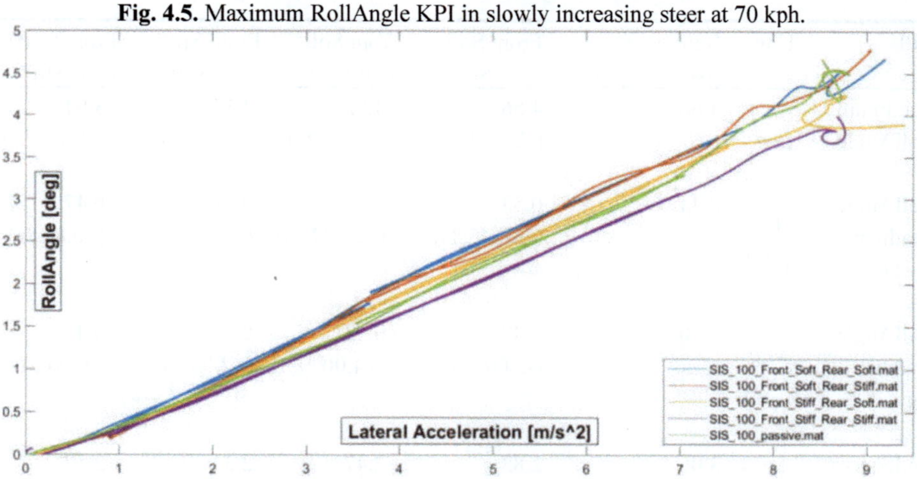

Fig. 4.6. Slowly increasing steer maneuver at 100 kph

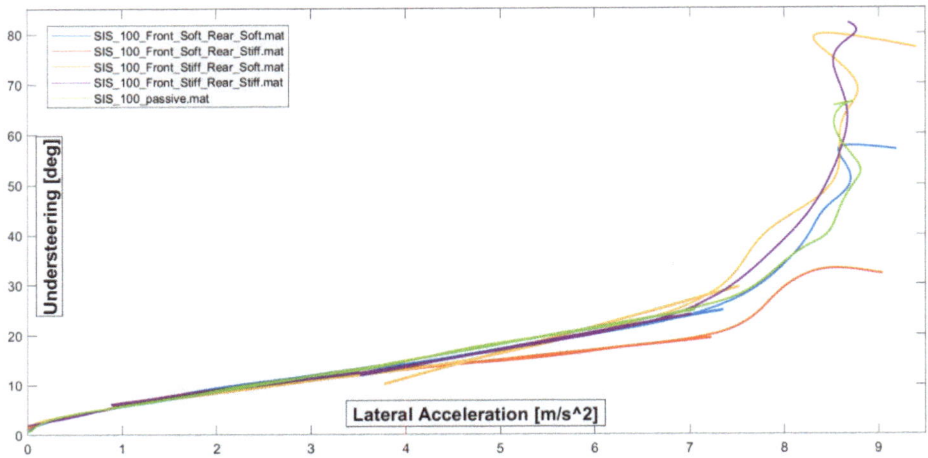

Fig. 4.7. Understeering during the slowly increasing steer maneuver at 100 kph.

Table 4.3. Slowly increasing steer KPIs at 100 kph.

KPI	Unit	Reference (passive system)	Front Soft Rear Soft	Front Soft Rear Stiff	Front Stiff Rear Soft	Front Stiff Rear Stiff
Maximum RollAngle	deg	4.66	4.56 (-2.10 %)	4.82 (+3.43 %)	4.37 (-6.22 %)	3.94 (-15.41 %)
RollAngle Gradient lowG	$\frac{deg}{\left(\frac{m}{s^2}\right)}$	0.44	0.53 (+20.45 %)	0.56 (+27.27 %)	0.51 (+15.91 %)	0.47 (+6.82 %)
RollAngle gradient highG	$\frac{deg}{\left(\frac{m}{s^2}\right)}$	0.50	0.49 (-2.00 %)	0.52 (+4.00 %)	0.51 (+2.00 %)	0.47 (-6.00 %)
Understeer gradient lowG	$\frac{deg}{\left(\frac{m}{s^2}\right)}$	3.06	2.83 (-7.52 %)	2.47 (-19.28 %)	2.32 (-24.18 %)	2.57 (-16.01 %)
Understeer gradient highG	$\frac{deg}{\left(\frac{m}{s^2}\right)}$	3.25	3.14 (-3.38 %)	1.95 (-40 %)	4.54 (+39.69 %)	3.59 (+10.46 %)

Figure 4.1 and Figure 4.6 illustrate the roll angle behavior during the slowly increasing steering maneuver, based on the resulting lateral acceleration. These figures highlight the differences in roll angle gradients as well as the maximum roll angle. The lines represent a fitted curve to give a clearer representation of noisy measurement data. Tables 4.1, 4.2 and 4.3 present the KPIs with a reference value, either from the passive system or, in case of the active system the normal driving mode. The tables provide a relative value in percent. To compare the active system (Figures 4.3 and 4.4, Table 4.2) with the semi-active system (Figures 4.1 and 4.2, Table 4.1) the normal mode is considered. The maneuver for the active suspension system is carried out in the same way as it is done with the semi-active anti-roll bars. With the semi-active anti-roll bars, the roll angle can be reduced by 11.3 % to a minimum of 3.6 degrees by adjusting the anti-roll bar on both axles to the stiffest possible setting. This results in a smaller roll angle than with the active system in normal driving mode (-7.8 %). By altering the stiffness of the anti-roll bars, the roll angle gradient can be influenced in both the low and high lateral acceleration areas. This gradient describes the amount of roll angle the vehicle produces per m/s^2 of lateral acceleration. According to Table 4.1 the range in which the roll angle gradient can be varied is a total 36.41 %. Compared to the active suspension (Table 4.2), the roll angle gradient is higher with the semi-active system. However, depending on the setting, the roll angle with the semi-active anti-roll bars is smaller. This is because the active suspension generates a higher gradient at the end of the lateral acceleration area (see Figure 4.3). In the various comfort modes of the active suspension, the vehicle actively leans into the turn, producing a roll angle gradient that is in the opposite direction of the normal mode. This is only achieved within a low lateral acceleration range. As the driver moves beyond this range, the roll angle begins to follow the normal behavior with a significantly higher rate starting from +200 % of the normal roll angle gradient in the high g range (see Table 4.2).

The understeer gradient quantifies the degree of a vehicle's understeering. Understeering is calculated by the difference of the SWA and the Ackermann angle for the driven radius. If the understeer gradient is positive, the vehicle is understeering, and if the gradient is negative, the vehicle is oversteering. A gradient of zero is referred to neutral steering. The stiffness of the anti-roll bars can influence this understeering behavior. Specifically, a stiffer front anti-roll bar results in more understeering and vice versa. This behavior is demonstrated in real tests by the semi-active system (as shown in Figures 4.2 and 4.7). A change in the understeer gradient can also be observed in the active system as depicted in Figure 4.4 and Table 4.2. The semi-active system, with a -12.9 % change, is not able to alter the understeer gradient in the low g range as much as the active system, which shows a -81.4 % change. However, the semi-active system can change the gradient in the high g range (-31.8 %) to a greater extent than the active system (+4.5 %). In Figure 4.7, during the maneuver at 100 kph with a setting of the front soft and rear stiff anti-roll bar, it is clearly noticeable that the understeer gradient in the high g range is reduced, followed by a change to oversteer at approximately 8.5 m/s^2 lateral acceleration. The trends of the maximum roll angle KPIs for the maneuver at 70 kph can be visually inspected in Figure 4.5.

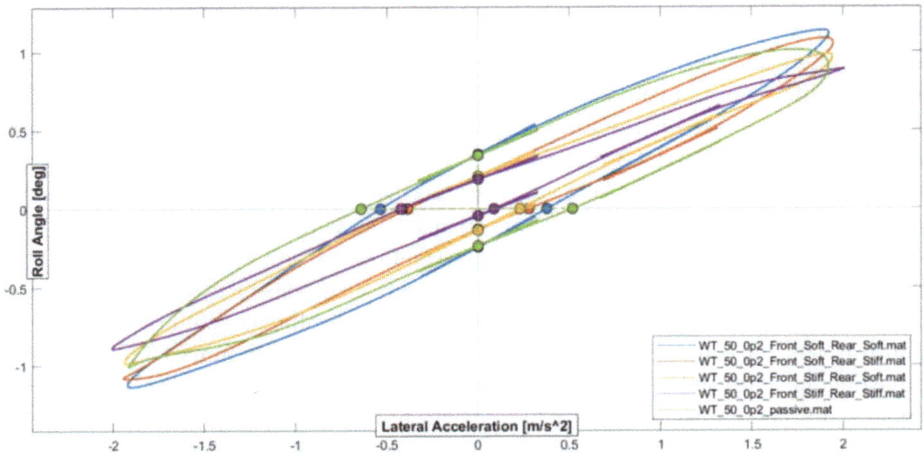

Fig. 4.8. Weave test at 50 kph with ay = 2 m/s^2.

Table 4.4. Weave test KPIs at 50 kph with ay = 2m/s^2.

KPI	Unit	Reference (passive system)	Front Soft Rear Soft	Front Soft Rear Stiff	Front Stiff Rear Soft	Front Stiff Rear Stiff
Maximum RollAngle	deg	1.01	1.14 (+12.87 %)	1.08 (+6.93 %)	0.98 (-2.97 %)	0.89 (-11.88 %)
RollAngle Gradient at ay = 0 m/s^2	$\frac{deg}{\left(\frac{m}{s^2}\right)}$	0.49	0.61 (+23.58 %)	0.51 (+2.85 %)	0.51 (+4.27 %)	0.45 (-8.94 %)
RollAngle Hysteresis	deg	0.57	0.59 (+2.09 %)	0.34 (-41.29 %)	0.34 (-40.59 %)	0.23 (-59.76 %)

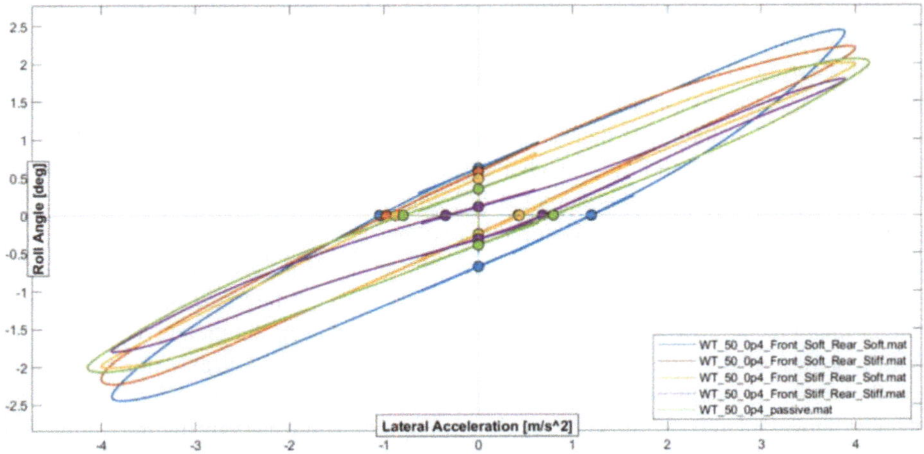

Fig. 4.9. Weave test at 50 kph with ay = 4 m/s^2.

Table 4.5. Weave test KPIs at 50 kph with ay = 4m/s^2.

KPI	Unit	Reference (passive system)	Front Soft Rear Soft	Front Soft Rear Stiff	Front Stiff Rear Soft	Front Stiff Rear Stiff
Maximum RollAngle	deg	2.06	2.44 (+18.45 %)	2.23 (+8.25 %)	2.01 (-2.43 %)	1.80 (-12.62 %)
RollAngle Gradient at ay = 0 m/s^2	$\frac{deg}{\left(\frac{m}{s^2}\right)}$	0.46	0.53 (+14.60 %)	0.58 (+25.71 %)	0.54 (+17.21 %)	0.36 (-21.13 %)
RollAngle Hysteresis	deg	0.74	1.3 (+75.91 %)	0.82 (+11.23%)	0.73 (-0.81 %)	0.43 (-42.49 %)

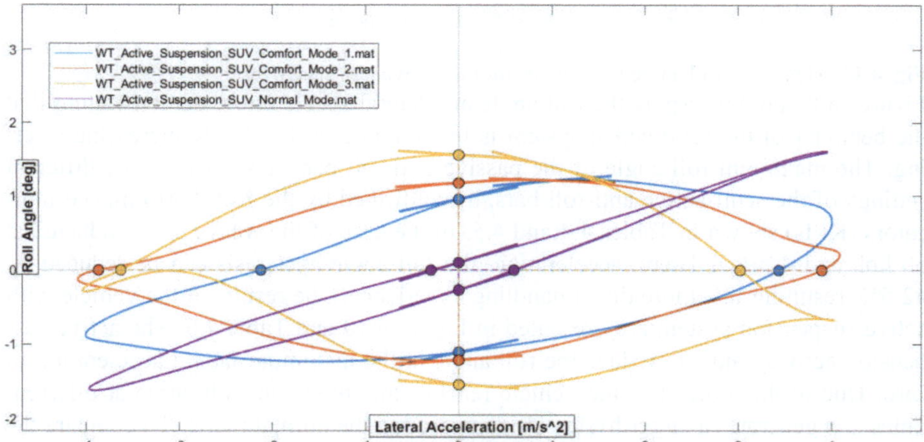

Fig. 4.10. Weave test with active suspension vehicle at 50 kph and ay = 4 m/s^2.

Table 4.6. Weave test KPIs with active suspension vehicle at 50 kph with ay = 4m/s^2.

KPI	Unit	Reference (normal mode)	Mode 1	Mode 2	Mode 3
Maximum RollAngle	deg	1.62	1.23 (-24.07 %)	1.22 (-24.69 %)	1.57 (-3.09 %)
RollAngle Gradient at ay = 0 m/s^2	$\frac{deg}{\left(\frac{m}{s^2}\right)}$	0.39	0.26 (-33.33 %)	0.08 (-79.49 %)	-0.08 (-120.51 %)
RollAngle Hysteresis	deg	0.37	2.07 (+459.46 %)	2.39 (+545.95 %)	3.10 (+737.84 %)

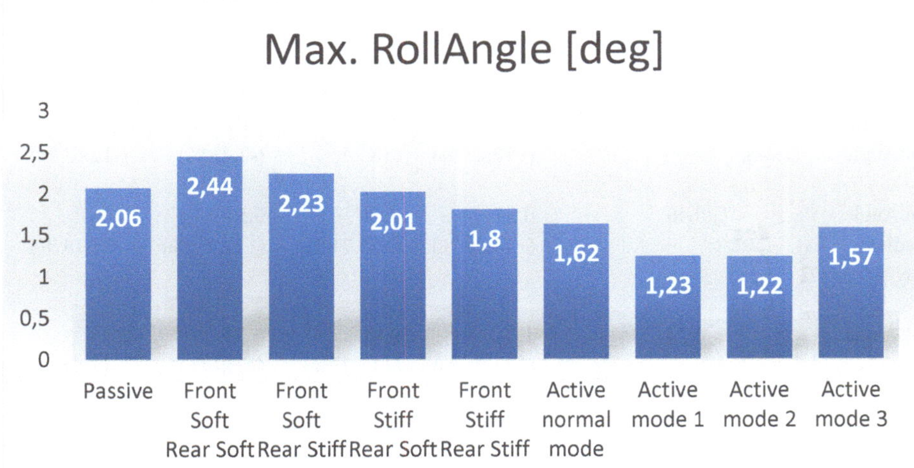

Fig. 4.11. Maximum roll angle KPI values for the weave test maneuver.

Figures 4.8 and 4.9 display the roll angle and lateral acceleration, demonstrating that the behavior of the semi-active system is like the one in the slowly increasing steering. The maximum roll angle of the passive anti-roll bars lies between the different settings of the semi-active anti-roll bars, as confirmed by the Key Performance Indicators (KPIs) shown in Tables 4.4 and 4.5. In the case of the weave test conducted at 50 kph and 4 m/s^2 lateral acceleration, the roll angle hysteresis can be reduced by 42.5%, resulting in a more direct handling experience. The results of the vehicle with active suspension system are presented in Figure 4.10 and Table 4.6. The active suspension actively aims to reduce the roll angle at the maximum lateral acceleration to zero. Due to this objective, the vehicle reaches the maximum roll angle at different times and generates a larger hysteresis compared to the normal mode. To compare the semi-active with the active suspension, the normal mode of the vehicle with active suspension is selected. Table 4.5 shows the roll angle with the front stiff and rear stiff settings at a maximum of 1.8 degrees. This is compared to the vehicle with active suspension (Table 4.6) which has a value of 1.62 degrees, a relative value of +11.1 %. For the roll angle gradient at zero lateral acceleration, the semi-active system has the lowest value with a stiff/stiff setting at 0.36 deg/(m/s^2), compared to the active suspension at 0.39 deg/(m/s^2), which is 7.7 % lower. The active vehicle has a smaller hysteresis value of 0.37 deg instead of 0.43 deg (+16.2 %) for the semi-active stiff/stiff setting. Figure 4.11 gives a visual overview the maximum roll angle values in the weave test.

5 Conclusion

The semi-active anti-roll bar, designed to match the torsional stiffness of traditional passive systems, not only maintains similar physical properties but also introduces variable stiffness and damping capabilities. This adaptability was reflected in the tests, where the semi-active system demonstrated a refined balance between comfort and stability, outperforming the passive system, particularly in dynamic scenarios.

Moreover, when compared to fully active systems, the semi-active technology exhibited comparable performance benefits in terms of vehicle dynamics reducing roll angles and moderating understeer gradients yet with lower energy consumption and system complexity. This positions the semi-active system as a viable, cost-effective solution that does not compromise on safety or functionality, evident from its fail-safe operation during electronic failures.

In conclusion, the integration of MR technology into semi-active anti-roll bars presents a promising avenue for future vehicle enhancements. It offers a strategic compromise, bridging the gap between passive simplicity and the high efficacy of active systems, making it especially suitable for electric vehicles due to the low energy consumption. The adaptability and efficiency of semi-active systems advocate for their broader application in automotive design, potentially setting a new standard in the industry for achieving optimal vehicular performance and passenger comfort.

References

1. Audi MediaCenter, https://www.audi-mediacenter.com/de/der-neue-audi-s8-begeisternde-performance-in-der-luxusklasse-2019-12256/die-fahrdynamik-12260, last accessed 2024/04/22.
2. Porsche Newsroom, https://newsroom.porsche.com/de/produkte/taycan/fahrwerk-18537.html, last accessed 2024/04/22.
3. ISO 13674-2:2016, Road vehicles - Test method for the quantification of on-centre handling.
4. ISO 13674-1:2016, Road vehicles - Test method for the quantification of on-centre handling.

Vehicle Simulation Methodology for MBD Process Considering Primary/Secondary Ride and Harshness

Taichi Fujita[1], Hiroyuki Tanaka[1], Katsuhiko Arai[1], Paolo di Carlo[2] and Marco Occhineri[2]

[1] Nissan Motor Co., Ltd.; 560-2 Okatsukoku, Atsugi City, Japan
[2] Siemens Digital Industries Software; Interleuvenlaan 68, Leuven, Belgium

taichi-fujita@mail.nissan.co.jp

Abstract. Ride comfort performance of commercial vehicle is one of the key factors for increasing its attractiveness, however because of increasingly complex systems, components and units, also the increasing vehicle weight due to vehicle electrification and intelligent driving systems, development becomes an increasing challenge and one that needs to be addressed up-front in the development cycle. In order to improve the comfort performance, it is important to consider not only general Ride Comfort (Primary/Secondary: up to 25z) but also Harshness(25-150Hz). To reduce the development costs which are caused by tougher requirements and enlarged performance scope, Model Based Design (MBD) techniques are increasing applied at an early development phase of the design. Application of their method in vehicle manufacturers has been conducted on ride comfort performance [1]. However, the scope in almost of all cases is only Primary/Secondary ride.
In this research, a hybrid system simulation model which can consider the full performance scope in early vehicle development phase, has been built and applied for vehicle evaluation. It is possible to consider high frequency phenomena that be contributed to Harshness by including a detailed suspension model, body transfer function (from Test or 3D-CAE result) as its characteristics. This model's effectiveness has been validated for performance study on early development phase. This paper shows the hybrid system simulation model building methodology and the result of a comparison between vehicle test and this model.

Keywords: Model Based Development, Ride Comfort, Harshness, Chassis Development, Performance Deployment.

P. E. Pfeffer (Ed.), *15th International Munich Chassis Symposium 2024*, Proceedings, https://doi.org/10.1007/978-3-662-71064-7_4

1 Introduction

In recent years, it becomes more critical to consider the vehicle process and architecture in early vehicle development phase. Vehicles are becoming increasingly complex. Electrification of vehicles and intelligent driver assist functions typically increase overall vehicle weight resulting in significant impact on dynamic performance. As this dynamic performance is a key aspect of the commercial appeal of vehicles it is necessary to perform detailed studies taking into account multiple attributes and combinations of vehicle functions. Such type of study would be unsustainable in terms of development time and costs without the application of simulation techniques.

In terms of vehicle comfort, sound and vibration in the cabin are also important factors. As key performance indicators (KPI), Ride Comfort and Harshness are used for instance. Generally, Ride Comfort performance has a scope of up-to 25Hz vibration phenomena, and these research areas have seen various previous studies using Model Based Development (MBD) by using a system simulation [1]. This research shows body and suspension considered as a rigid object with the sprung and un-sprung resonance also included in the modeling approach. This modeling method is widely known in the research of Ride Comfort performance as the system simulation and one of its biggest advantage is the easiness in functions and factors parameterization. By analyzing the research studies and development cases, the system simulation modeling method is applied in the Primary and Secondary ride comfort area below 25Hz, however phenomena in the range of 25Hz to 150Hz are considered as necessary to characterize the comfort performance of the vehicle [2], and so the performance development are extended typically to this Harshness region.

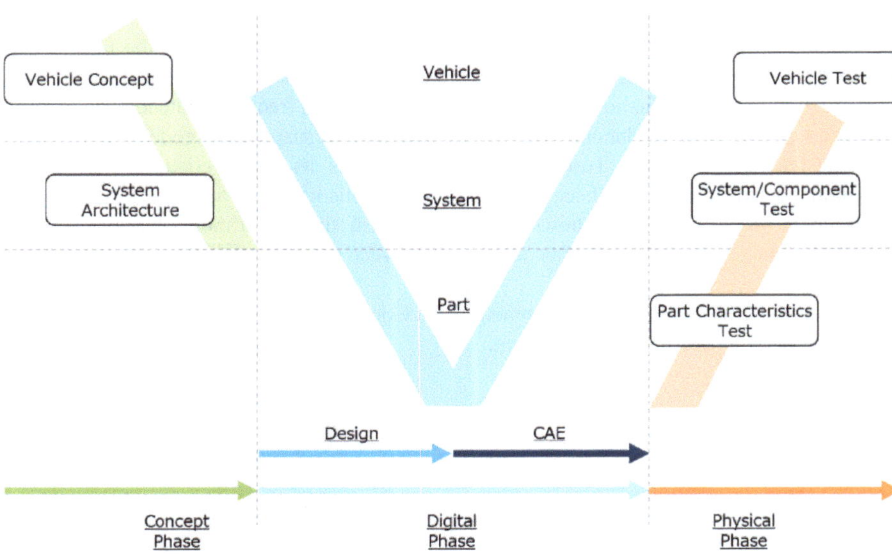

Fig. 1. The general vehicle development process and functional deployment from concept phase to the physical phase

Focusing on the vehicle development process (to see Fig.1), these middle frequency phenomena are difficult for planning or designing in early development phase, since vehicle level sound and vibration are related to complicated phenomena such as body and suspension elastic resonance. In order to address this problem, the simulation process needs to more precisely reproduce the vibration phenomena. The key technology will be not only low Degree Of Freedom (DOF) but also including detailed geometry and dynamic characteristics. If these middle frequency phenomena can be considered, frontloading the design at early stage of development can reduce the likelihood that a performance of vehicle will be rejected on physical vehicle experiments in the late stage of development saving both prototype hardware cost and development time.
In this paper, a new method applying a hybrid of system simulation, multi body dynamics and experiment data, is proposed to design the vehicle Ride Comfort / Harshness performance against target KPIs at both system and component level.

2 Concept of a New Simulation Methodology for Applying Ride Comfort and Harshness

Before mentioning the modeling and new method for predicting the vehicle KPIs, All the key factors for reproducing the sound and vibration phenomena are stated in this section. By the conventional experiment, the components, systems or indicators which can contribute to the vehicle Ride Comfort performance are listed up in order to apply them in CAE modeling.

2.1 Contributed Phenomena for Ride Comfort and Harshness

The key factor contributed to the vehicle level can be divided into these areas as shown in Fig.2 derived from the traditional experience with vehicle test and 3D-CAE simulation. One is the body vibration characteristic, included in this are the rigid body modes (resonance modes on spring), trim resonance modes of vehicle, and other low frequency resonances such as those relating to the doors These are needed to consider as one of the key factors up-to 150Hz. The body system simulation approach is more and more attracting attention nowadays, however most research only mention the body-frame system simulation [3]. Therefore, it is quite difficult to consider the complicated resonance including middle range of frequency. In this research, the body transfer characteristics themselves are considered into the CAE model to reproduce the detailed vibration or sound characteristics.
In the following, the suspension characteristics are focused on. Based on conventional experience the related factors are rigid mode of suspension links and tire resonance. Once focusing on the frequency up-to 150Hz, it is thought that by modeling rigid suspension link, spring and damper, the phenomena could be roughly grasped in the target frequency range. This research adopts multi-body dynamics to solve these suspension mechanisms. Furthermore, it is widely known for the tire and its resonance to significantly contribute the vehicle vibration [4]. As general Ride Comfort performance (only Primary and Secondary Ride), the tire vertical / lateral / longitudinal static stiffness, the

first tortional resonance and the first vertical bounce resonance are the main contributing factors. Regarding tire vertical static stiffness, it appears as coupling with the suspension, and this affects not only the un-sprung resonance but also vehicle vibration like as the floor vibration. The first tire tortional and vertical bounce also have similar impacts for the vehicle level, moreover these exist only in one resonance without coupling the DOF. The resonance as mentioned above are also important factors for planning and designing the vehicle performance.

The aim of this research is to build a CAE model in order to design and deploy the key factors for Primary, Secondary Ride and Harshness in both early development phase and planning phase before vehicle experiment. The hybrid method between system simulation, multi body dynamics and characteristics obtained by experiment is essential to achieve the aim as above. This proposed method considered a compromise of obtaining a sufficient accuracy of reproducing each phenomenon and also the ease of use in performance planning considering also the deployment of those models.

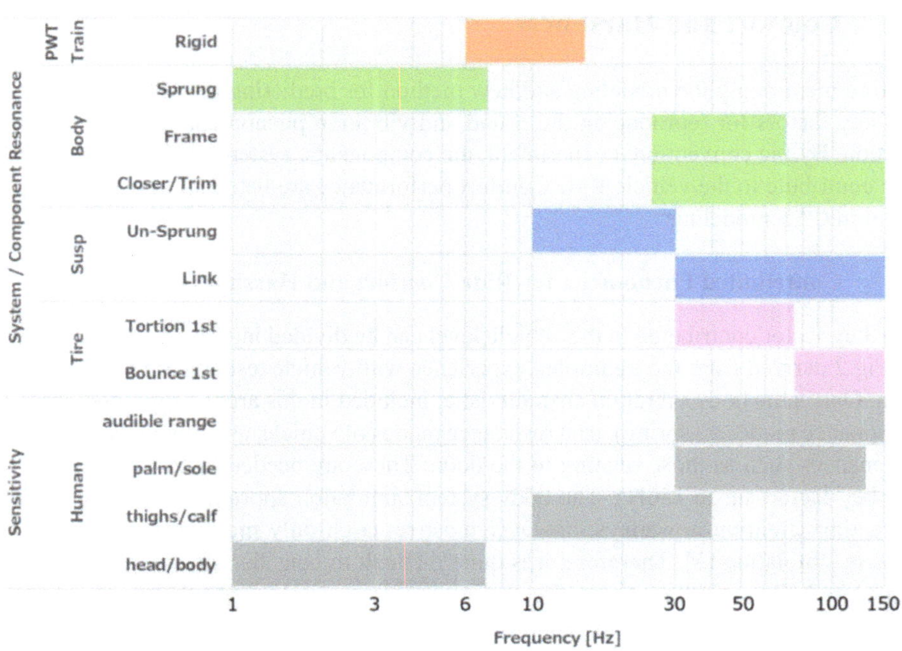

Fig. 2. The key factors contributing to vehicle level sound and vibration with respect to frequency range.

2.2 Detailed Concept for the Connection of Each System

The new model and simulation concept is proposed based on the phenomena as mentioned above. Fig.3 shows the details of the model sketch and the contents are as follows.

- Body: embedded in as transfer function and obtained from test or 3D-CAE
- Power Train: mass-spring-damper system with 6 DOFs
- Suspension: multi body dynamics with rigid link, spring, and damper
- Tire: dynamic tire model which can reproduce the correct vibration modes
- Road: 2D road surface with respect to above tire model

Body transfer characteristics and road surface are applied by measured data. Other parts are defined as system simulations or multi body dynamics to solve the kinematics and dynamics theory. Power train model is also adopted with six DOF model. Moreover, power train mounting especially hydro bushing, are composed by multiple springs and masses for expressing its characteristics that include amplitude and frequency dependency. In most cases, vehicles equip power train hydro mounting, therefore this modeling is also essential for the simulation. As for the suspension, the rigid mode of each suspension links has high contributed to the dynamics, in view of this, a detailed multi body dynamics model including the geometry and dynamic stiffness and damping are adopted. Tire model is required for reproducing the phenomena caused by tire to guarantee the vehicle level prediction accuracy. In the vehicle early development phase, tire specification is one of the key factors to decide considering Ride Comfort, Handling, Energy Assumption (Rolling Resistance Coefficient), NVH, cost, layout, etc. The tire model used has a specification examination function. Its function is valid for selecting tire specification that consider the multiple performance. In addition to these, road surface input is compatible with tire and able to simulate same condition as vehicle testing.

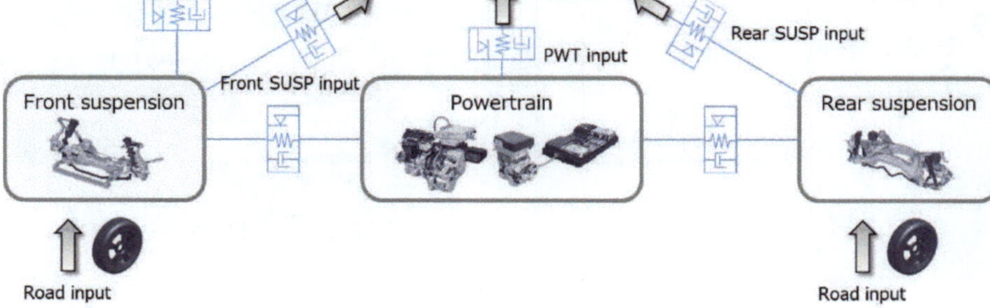

Fig. 3. System simulation and model concept considering both Primary/Secondary Ride and Harshness

3 Process and simulation overview

Vehicle Ride Comfort performances depend on excited frequency range. Different excitations generated by different road profiles or systems (engine, driveline, etc.) are able to highlight resonance of vehicle components at specific frequency range. The Ride Comfort area is usually divided in three main sub-ranges identified as follows:

- Primary Ride: car performances up to maximum 5-6Hz
- Secondary Ride: car performances from 5Hz up to maximum 25Hz
- Harshness: car performances from 25Hz up to maximum 150Hz

The Primary Ride phenomena focuses on the rigid body modes of the vehicle upper body sprung masses (e.g. Pitch, Roll, etc.) and the related road profiles consisting of asphalt disturbances with long wavelengths but relevant divergence in vertical (z) direction from flat condition.

The Secondary Ride phenomena, on the contrary, mainly involves the vehicle chassis un-sprung masses dynamic behavior together with engine rigid mode coupling and the components structural flexibilities start becoming relevant.

The Harshness range deeply involves the vehicle components structural flexibilities while chassis un-sprung mass dynamic remains relevant. If the vibration phenomena are particularly strong, the term "boom" is often used (boundary vibration-acoustic).

3.1 Simulation for Low Frequency

Ride Comfort is key performance when chassis should be designed and developed, since the asphalt asperity filtration capability defines the level of vibration, and therefore comfort or dis-comfort, at interfaces between vehicle and human body.

Leveraging on a MBD approach, the physical modeling of the system can be initiated earlier in the design process to assess future product performances and verify design choices. This allows to frontload suspension design at early stage of the chassis development making the overall process more efficient and optimized. Fig.4 shows the MBD area of application during development V-Cycle.

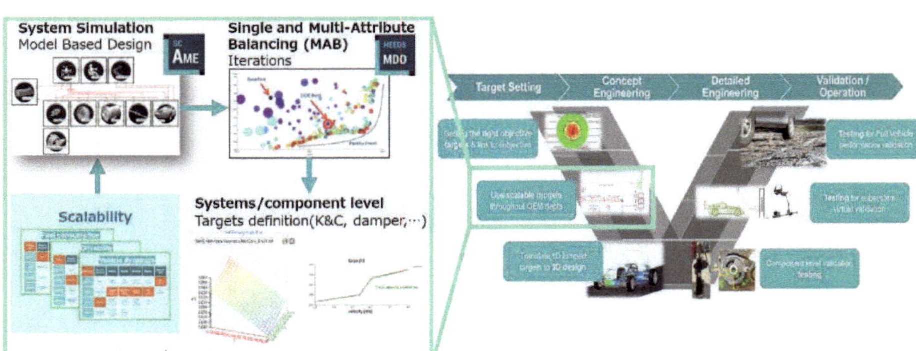

Fig. 4. Model Based Design & System Simulation application at early design stage through V-Cycle.

MBD approach is based on a combination of system simulation [5] and model scalability based on which the model complexity depends on the performance to be simulated without affecting the accuracy of the prediction as showed on Fig.5. Model scalability enables the Ride Comfort study during initial stage when less data is available amplifying the area of investigation and suspension virtual tuning effectiveness.
In addition to scalability, any methodology able to extend predicted frequency range of system simulation based models, is crucial to anticipate suspension optimal design and solve issues otherwise difficult or more expensive to be eliminated during advanced development stage.

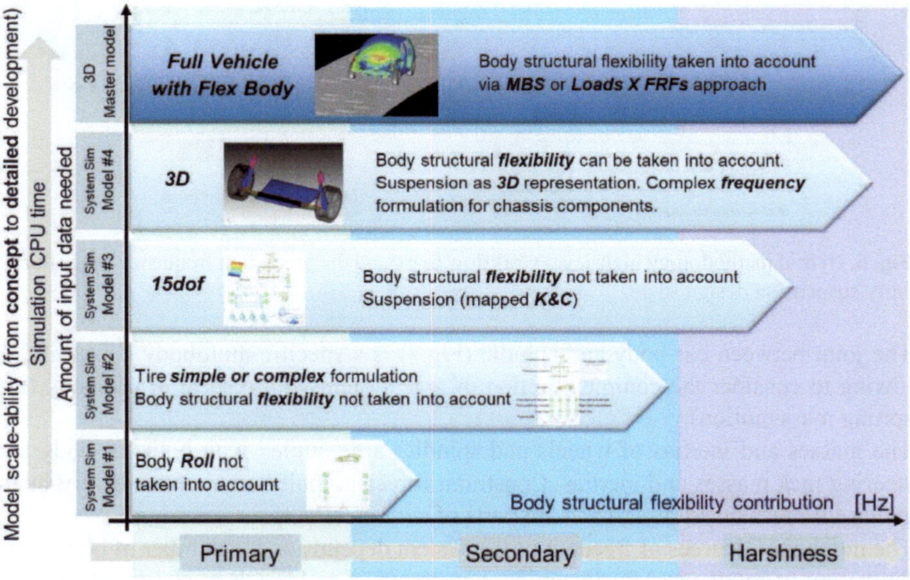

Fig. 5. System Simulation Scalability overview for Ride Comfort – start to design suspensions from early development stage with low complexity but already high level of prediction.

3.2 Combination of Low DOF Model and Suspension/Body-characteristics

In the current paper, a hybrid methodology (Fig.6) to use FRFs from either test or CAE (VTFs and/or NTFs) in combination with Loads from simulation to predict vibration and/or noise at target points is presented. This allows to predict seat rail vibration using a system simulation model (typically limited in terms of max frequency) during cleat impact (Harshness) up to 150Hz.
System simulation (Simcenter Amesim) full vehicle model deployed for the Harshness study is based on native 15DOF model enhanced with additional super-component able to properly model the suspension longitudinal dynamic behavior which plays, along with vertical one, a relevant role during the cleat impact maneuver.
The core of 15DOF model is the chassis module to which all car submodules can be connected (suspension - spring, damper, end stop, antiroll bar - aerodynamic module, tire, road, sensors, engine, brakes, steering system, powertrain module, external

load,...). Chassis module is a multibody system containing several pieces, "car-body", "steering rack", "spindle (un-sprung non-rotating mass)", "wheel" and all the mechanical joints between these elements.

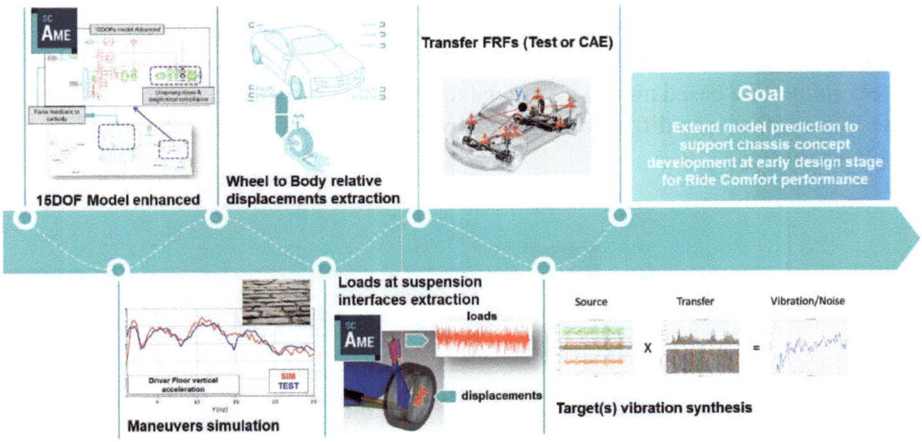

Fig. 6. Hybrid methodology high-level workflow to extend the prediction frequency range during early suspension design stage.

The joint between car-body and spindle (Fig.7) is a specific multibody constraint allowing to consider the complex motion of axle system with respect of car-body (unsprung mass motion).
The masses and inertias of wheels and spindles are coupled with both car-body and steering rack masses and inertias. Constraint forces in multibody joints are very interesting for mechanical design and are parts of car model outputs.
The number of degrees of freedom of the system depends on the number of bodies and the nature of joints used for kinematic constraints. For 15DOF model the degrees of freedom are defined as following:

- Car-body: 6 DOFs
- Steering rack body: 1 DOFs
- 4 spindle bodies: 4 DOFs
- 4 wheel body: 4 DOFs

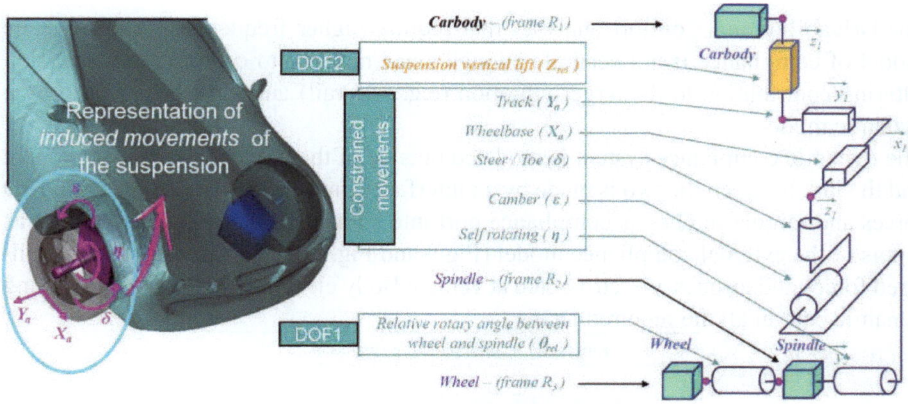

Fig. 7. 15DOF model: Schematic representation of the joints chain from wheel to car-body

The chassis model also allows the modeling of suspension kinematics and compliance also called "elasto-kinematics". Under load, the compliance caused by the elastic characteristics of the suspension bushings, modifies the rim plane position and orientation. Suspension compliance differs from suspension kinematics which is the wheel constrained motion due to the suspension geometry. Both kinematics and compliance are present and superimposed during vehicle maneuver and play key role into suspension performance definition (Fig.8).

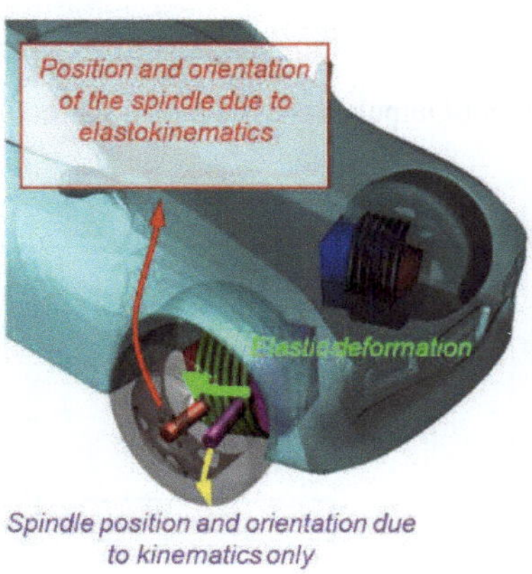

- Superimposition of kinematics motions and elastic deformations for translations & rotation:

$$d_{rel\ ELASTOKIN} = d_{rel\ ELAS} + d_{rel\ CIN} :$$

$$\begin{cases} x_{rel\ ELASTOKIN} = x_{rel\ ELAS} + x_{rel\ CIN} \\ y_{rel\ ELASTOKIN} = y_{rel\ ELAS} + y_{rel\ CIN} \end{cases}$$

$$r_{rel\ ELASTOKIN} = r_{rel\ ELAS} + r_{rel\ CIN} :$$

$$\begin{cases} \delta_{rel\ ELASTOKIN} = \delta_{rel\ ELAS} + \delta_{rel\ CIN} \\ \varepsilon_{rel\ ELASTOKIN} = \varepsilon_{rel\ ELAS} + \varepsilon_{rel\ CIN} \\ \eta_{rel\ ELASTOKIN} = \eta_{rel\ ELAS} + \eta_{rel\ CIN} \end{cases}$$

Fig. 8. 15DOF model: total wheel position and orientation as superimposition of kinematics and elastics contributions

The “standard” 15DOF model only considers the suspension vertical as dynamic mode, whereas the rest of the wheel DOFs are defined statically in terms of position/rotation making this representation suitable for Primary Ride and Handling analyses that are focused on low frequency range 1-5 Hz.

For Drivability and Comfort analyses that require higher frequencies, a quasi-static model of compliance is not accurate enough since not able to capture the suspension filtering contribution to the target vibration (e.g. seat rail) and must be replaced by a dynamic model.
The dynamic compliance system is modeled outside of the 15DOF chassis component and the link between the two is made by an interface component which converts the 3D forces and torques at chassis compliance port into signal variables that can be used as inputs of the external compliance model (Fig.9 and Fig.10). The suspension output filtered forces and torques are retro-acted at vehicle Body element enabling the un-sprung dynamic path in all the required DOFs.

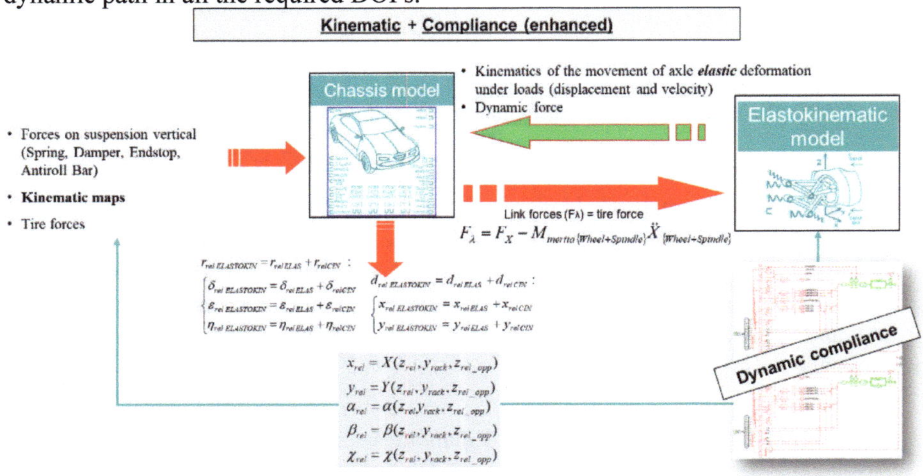

Fig. 9. 15DOF model: schematic forces-displacements Input-Output to chassis model representation and elasto-kinematics contribution

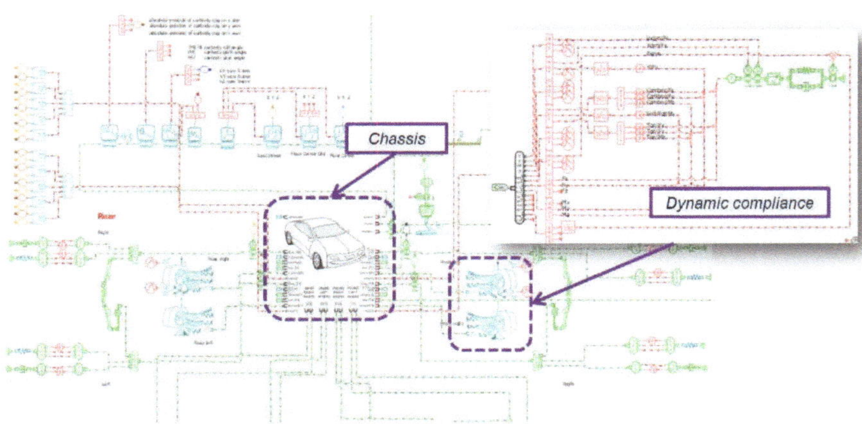

Fig. 10. 15DOF model: enhancement with Dynamic stiffness implementation to enable longitudinal mode contribution

The enhanced dynamic compliance model is highly scalable at both: suspension representation and DOFs level:

- From simple mass, spring, damper to more complex frequency dependent formulation (including hydro-bushing)
- From single longitudinal DOFs to many (e.g. lateral, rotation) including cross coupling as showed in Fig.11

This modeling flexibility allows to extend the prediction accuracy of System Simulation vehicle model by involving additional performances into the study and extending the validity of the frequency range.

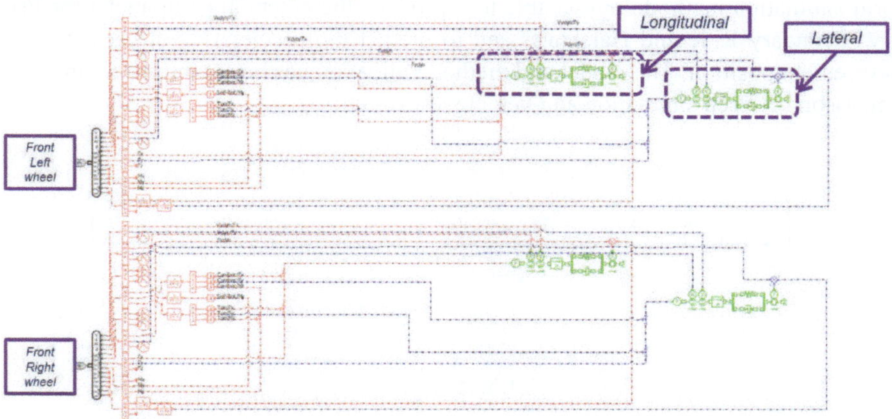

Fig. 11. Dynamic compliance can be extended to all wheel DOFs illustrated in the figure by adding the lateral contribution as well

4 Simulation Results and Validation with Vehicle Experiment

The simulation methodology and hybrid model can predict vehicle level vibration or sound. The vibration result below 30Hz predicted by 15DOF Advanced (Adv) model is shown to begin with. In addition to this, the comparative verifications between hybrid simulation results and vehicle test results are carried out in the scope of up-to 150Hz.

4.1 Comparison between Simulation and Test

Firstly, a simulation result that is the basis for the comparison of 15DOF Adv model is shown in Fig.12. This simulation result shows the vibration level running on uneven-road and includes 6Hz to 25Hz phenomena. As shown in the figure, the gain level has reasonable accuracy compared to the vehicle test data, and also this simulation can reproduce powertrain resonance (around 6Hz), un-sprung resonance (below 20Hz) and their peak gain. As this result, at the region in Primary-Secondary Ride, the proposed system simulation model can be utilized to deploy the vehicle level, system component or functional design.

Additionally, the result from the proposed hybrid methodology evaluation is in Fig.13. In this case, the simulation setting is on the crack-road at 40km/h and evaluation point is the floor vertical vibration (acceleration).
By adopting 15DOF Adv model, sprung, PWT and un-sprung phenomena are predicted up-to 25Hz as same as Fig.12. In this frequency range, a similar level of accuracy with the hybrid method simulation is achieved. For the range of 25Hz to 150Hz, the peaks caused by the suspension rigid resonance mode can be observed and tire 1st bounce mode is also seen around 90Hz. Comparison between only 15DOF Adv and hybrid simulation improves overall gain level from 25Hz to 150Hz. These results show the hybrid simulation methodology is able to reproduce the essential phenomena for Primary/Secondary Ride and Harshness and to predict the acceleration at each vehicle measurement position. The model and method validation has been completed, through comparison between simulation and vehicle test results.

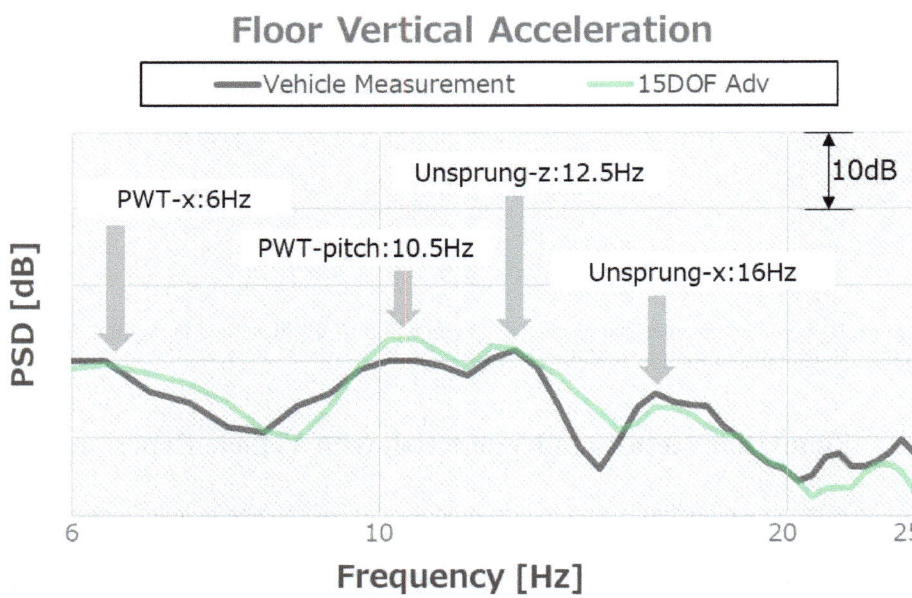

Fig. 12. Floor vertical vibration comparison between vehicle test and system simulation result at uneven road

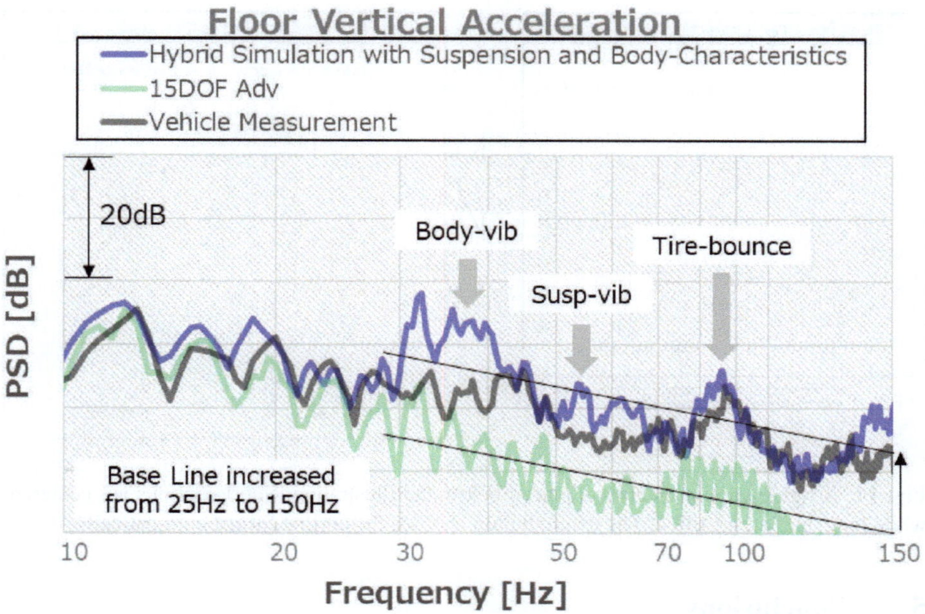

Fig. 13. Floor vertical vibration comparison between 15DOF Adv and hybrid simulation result on cracked road

4.2 Contribution of Part Characteristics Studying

The comparison of simulation and test confirms the application of the hybrid method for assessing vehicle comfort performance at a nominal set of known parameters. To apply this method in design deployment, is it necessary to validate the approach from perspective of design parameter changes, and this is presented in this subsection.

Designing the vehicle damper characteristics are described from here, as one case. In the early development phase, designing and planning damping characteristics of a shock absorber considering the vehicle mass and sprung resonance (0.1Hz to 6Hz) are highly significant for the car and its performance [6]. This parameter influences all of the vehicle KPIs, and is an essential tuning parameters for improvement of Ride Comfort.

The effect of a simple tuning of the damper force is shown in Fig.14. Here, the result of tuning the damper was compared with the measured result, and it was confirmed whether the phenomenon could be relatively reproduced. The results show that the reduction margin of certain peaks can be verified relatively in the frequency band above 30 Hz.

From these results, there is some confidence that the simulation method in this case was effective for both planning the vehicle design values in the early phase and confirming the vehicle achievement in the late phase.

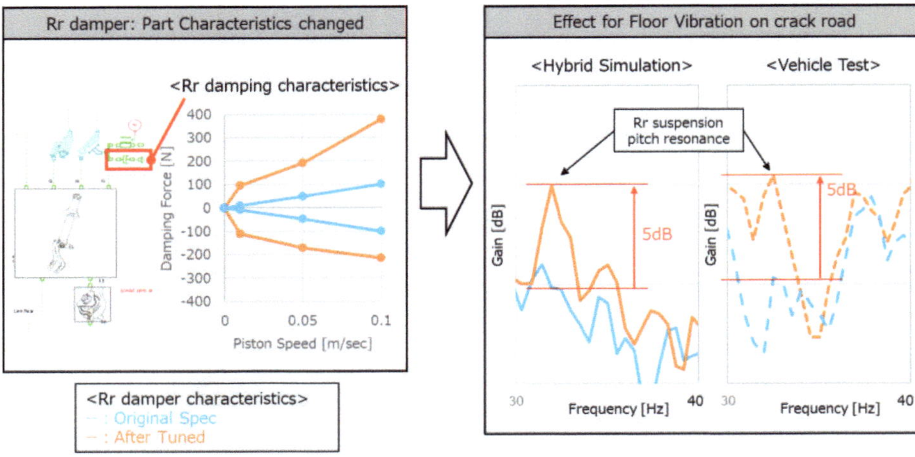

Fig. 14. A figure caption is always placed below the illustration. Short captions are centered, while long ones are justified. The macro button chooses the correct format automatically.

5 Conclusions

In this paper, we developed a vehicle prediction method that is a hybrid of system simulation, multi body dynamics, and experimental results that can be used in both the early and late phases of vehicle development. The 15DOF Adv model can include unsprung resonance, vertical / lateral / longitudinal. This modeling is able to predict and study the kinematics and compliance related to up-to 25Hz. Coupling with it, detail suspension model and body vibration transfer function, this proposed method can reproduce more high frequency phenomena: suspension rigid mode, trimmed-body detail mode. This hybrid simulation method shows vehicle level vibration and sound prediction can be conducted considering the higher frequency phenomena of Harshness.
Using the developed method, the vibration level is consistent with the actual measurement result and can be predicted as an absolute value. If only low frequency (up-to 25Hz) is required, a 15DOF Adv model can be utilized. On the contrary, when the vibration which includes higher frequency range up-to 150Hz is necessary, the hybrid simulation method with respect to 15DOF Adv + multibody suspension + VTF/NTFs is valid for both Ride Comfort and Harshness. In addition to this, it was shown that the effect in changing design and planning parameters could be comparatively evaluated.
By using this method, it has become possible to design and evaluate the unique Riding Comfort performance (Primary + Secondary + Harshness) across the vehicle design process, and its application will be advanced and extended to other attributes.

References

[1] I.-S. Jeong and V. Grange, "Target setting of Driving Performance index for Module-based Architecture development of Product family," *ChassisTech Conference,* 2023.

[2] T. Hiroyuki, A. Satomura and T. Suma, "Prediction and Analysis Technology Development for Impact Noise," *SAE International Journal,* 2014.

[3] R. Kakehashi, T. Yamazaki, K. Kurihara and H. Kawai, "Analysis by energy transmissibility using a simplified model considering idling vibration and engine shake," *Transactions of the JSME,* vol. 89, no. 924, 2023.

[4] F. Calabrese, L. Dusini, M. Bäcker and A. Gallrein, "Managing the variety of potential tire / wheel sizes in the early vehicle development process," *In Pfeffer P. (eds) 9th International Munich Chassis Symposium,* pp. 827-847, 2018.

[5] S. BARALE, A. PLISSON, J. GUILLET, J. LAGNIER and M. ALIRAND, "Improved Functional Modelling in Comfort," *ChassisTech Conference,* 2013.

[6] S. Barale, P. Vars, J. Guillet, M. Alirand, J. Lagnier , P. Toulemont, "COMFORT ANALYSES OF THE HYDRACTIVE SUSPENSION USING A DRIVING SIMULATOR," *Driving Simulation Conference,* 2014.

From customer requirement to individual contact points within rotation-translation-converters

Dipl.-Ing. (FH) Bernhard Wiesneth and Dr.-Ing. Simon Merz

Schaeffler Technologies AG & Co. KG, Industriestraße 1-3, 91074 Herzogenaurach, Germany

Abstract. The transformation of rotational to translational movements in the vehicle is carried out by rot-trans-converters. For example, planetary roller gear (PRG) and ball screw drives (BSD) impress with their robust design and high power density.
For more than 20 years, Schaeffler has been developing rot-trans-converters for electromechanical steering systems, where these high-precision components perform their duties with low friction and low noise. In rear-axle steering systems, PRGs are used due to their higher efficiency and good self-locking behavior compared to traditional trapezoidal screws.
BSDs also continue to impress in braking technology: they are used in particular for electromechanical parking brakes and brake boosters to ensure fast and effective braking performance of the vehicle. In addition to braking and steering applications, numerous BSD and PRG product variants are in series production, e.g. for clutch actuators and exhaust gas recirculation systems.
As an example, the selection and the design of the suitable rot-trans-converters will be shown based on customer requirements. Insights for more detailed analytical methods are applied depending on the selected subsystem.
Beside quasi-static methods, which range from system observation with an elastic environment to contact pressure of the single ball contacts, the focus is also on multi-body dynamics, such as vibration analysis of rolling elements or the influence of shear forces on the rolling element contact of the respective subsystem.
In the following, the authors describe the advantages of selected rot-trans-converters and offer an outlook on new and innovative fields of application, such as electromechanical service brakes and Ride Height Adjustment in vehicles.

Keywords: rot-trans-converter, ball screw drives, planetary roller gear, design, simulation.

1 Introduction

1.1 Rot-trans-gearboxes in the automotive industry

With the development of Rudolf Benz's first motorized car in 1886, the functional principle of converting linear motion into rotational motion and vice versa was implemented for the first time for an automobile. During the following approximately

P. E. Pfeffer (Ed.), *15th International Munich Chassis Symposium 2024*,
Proceedings, https://doi.org/10.1007/978-3-662-71064-7_5

140-year history of the automotive industry, numerous other innovations based on this operating principle were developed and to a large extent implemented in series production. The fields of applications can be found in almost all areas of automotive engineering (e.g. chassis, powertrain, and interior). The innovative strength for these developments is found in the increase in comfort and efficiency, but also in the optimization of active and passive safety.

1.2 Rot-Trans-Converters current applications

In the 1990s, intensive development of electromechanical brakes (EMB) was carried out for the next generation of brake systems of modern passenger cars. The implementation of these systems into mass production took more time with the first applications launching approximately 30 years later. A previous variant of the electromechanical brake with lower requirements was the electric parking brake (EPB) which was integrated within the hydraulic brake calipers of the rear axle. This actuator consists of a motor/transmission unit combined with, either a trapezoidal screw, ball screw or ball/ramp principle.
In advance to the implementation of EPB, a wide variety of series applications for other Rot-Trans converters have found their way into automotive series production (see figure 1). For example, the first purely electromechanical steering systems (EPS) were established in the year 1988 [1] by means of a worm gear in a column driven power steering system. Later, pinion and ball screw drives would be introduced in EPS systems and are standard technology today. In the field of electronic brake systems several e-boosters using rack-and-pinion, trapezoidal screws and ball screws drives were also brought to series maturity.

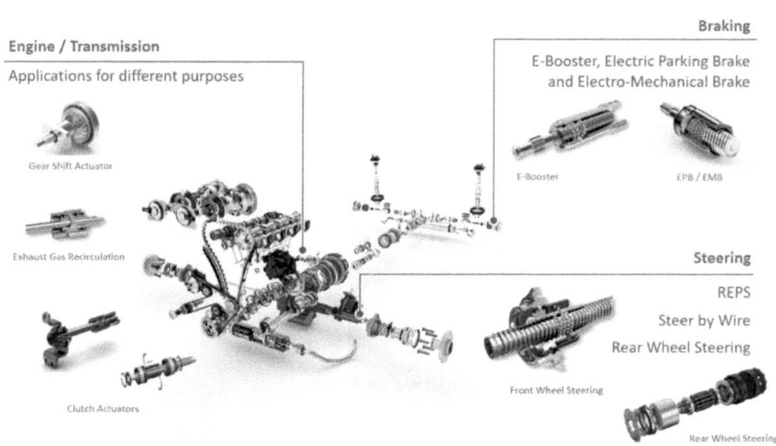

Fig. 1. Examples for Rot-Trans-Converter application fields

Rot-trans-converters also made their way into series production in dual-clutch transmissions and state-of-the-art rear-axle steering. In addition to well established ball or lead screw drives, planetary roller gears (PRG) and synchronized planetary roller gears (SPRG) were used for the first time.

Fig. 2. Planetary roller gear assembly

2. General customer requirements

Detailed customer specifications can be classified into following groups of requirements:

- Quality
- Dimensional
- Functional
- Lifetime
- General and legal requirements
- ...

2.1 Key requirements for the selection of rot-trans-converters

To select the proper rot-trans-converter, a trade-off between system costs and overall system functionality is essential. Table 1 lists examples of requirements and evaluations which helps to select the appropriate converter for specific application.

	BSD ball screw drive	PRG planetary roller gear	Rack and Pinion Type	TRG trapezoidal screw	Cam Shaft	Ball Ramp

efficiency	++ (b)	+ (c)	++ (b)	--	0	++
Self-locking	--	0	--	++ (a)	0	--
friction	++	-	++	--	--	--
load capacity	0	++	0	++	+	-
wear vs. lifetime	++	++	+	--	-	++
slippage	++	-1)	++	++	++	0
stroke	++	++	++	++	-	--
pitch	0	++	-	+	--	--

1) No slippage with synchronized PRG

Table 1: Selection criteria for Rot-Trans-Converters

Depending on who is doing the assessment and which application is considered the assessment may differ. In some application, certain requirements can be so critical that several Rot-Trans Converter elements can be excluded. Three examples will be outlined to illustrate this scenario:

a) Looking at the evaluation of the trapezoidal screw, it is highly rated in a variety of requirements. In addition, the impressively simple design of the trapezoidal screw leads to almost unbeatable cost advantages, as long as the disadvantages of low efficiency, high friction and the resulting wear can be managed. This type of machine element is often used for parking brake applications as well as interior application like steering column adjustments.

b) If efficiency and friction are of the utmost importance for the application, there is no getting around implementing a ball screw or rack-and-pinion gear. This is highlighted by their use in applications such as front steering gears and electro-mechanical brake boosters.

c) If, within certain limits, self-locking is required while maintaining high efficiencies, converters such as the planetary roller gear can score. The advantage of this rot-trans-converters is that it can be tuned in such a way that the PRG remains in a certain end position without applying a holding current to the actuator. In combination with its relatively high overall gear ratio, it may also be possible to eliminate an additional gear stage which might be required using an alternative rot-trans-converters. Taking these positive properties into account combined with the conceptually induced low wear compared to trapezoidal screws, the choice is often in favor of the PRG. Despite the higher procurement prices of this gear, it is often possible to achieve an optimal "design to costs" ratio.

3. Specific system requirements and the influence on single contact conditions

Based on selected examples, specific system requirements are broken down to the individual rolling contact of the respective assemblies and the resulting consequences are presented.

3.1 Brake application Influence of radial forces and orientation of ball nut to the ball contact pressure (hertzian pressure)

It is becoming increasingly important to avoid time-consuming development loops for the development of new, promising fields of application especially with often time critical milestones from initial inquiry to final SOP. Modern CAE tools help to set the course in good time, provided that the application is sufficiently described and can be incorporated into the definition of a suitable design at an early stage of development.

3.1.1 Influence of radial force of a ball screw assembly to the ball contact

To put it simply, the function of a disc brake can be described as transmitting the braking force via a piston to the brake pads and thus to the brake disc, and by means of friction between the pads and the rotating disc, the vehicle is braked. The resulting braking torque is transmitted from the brake disc via the friction pads into the brake caliper, whereby part of the torque must also be supported by the braking force generating machine element e.g. the ball screw.

The first Bearinx® calculation shows the influence on the highest loaded ball contact with and without radial load (see figure 3). Depending on the considered load case the hertzian pressure (pH) will rise significantly and may lead in worst case conditions to the situation that the ball screw design need to be modified because it is insufficiently dimensioned (see figure 4).

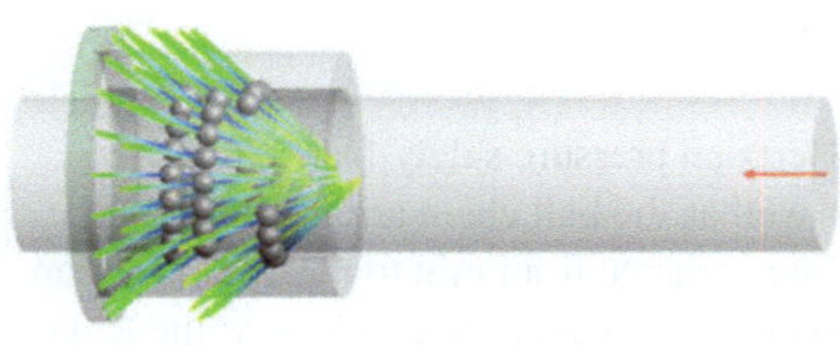

Stroke in mm	Fax in N	Frad in N	pHi in MPa	pHo in MPa
1	1000	0	1976	1808
2	2000	0	2460	2253
3	3000	0	2792	2559
4	4000	0	3051	2798
5	5000	0	3268	2998
6	6000	0	3455	3171
7	7000	0	3621	3325
8	8000	0	3771	3464
9	9000	0	3908	3591
10	10000	0	4034	3709

Fig. 3. Hertzian contact pressure for 10 kN axial force without radial force

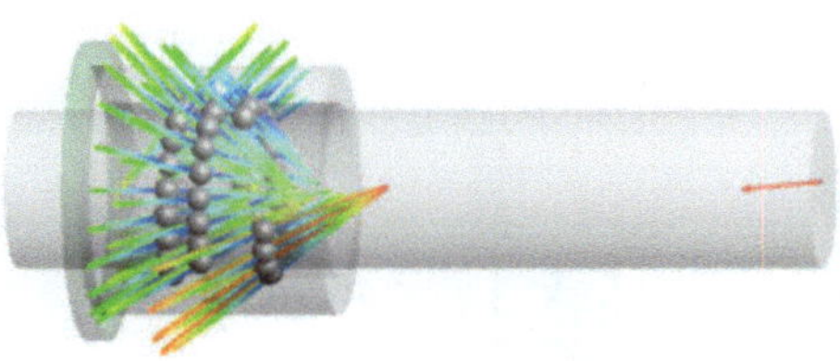

Stroke in mm	Fax in N	Frad in N	pHi in MPa	pHo in MPa
1	1000	100	2741	2516
2	2000	200	3335	3067
3	3000	300	3746	3449
4	4000	400	4075	3756
5	5000	500	4354	4016
6	6000	600	4599	4246
7	7000	700	4820	4453
8	8000	800	5022	4644
9	9000	900	5210	4820
10	10000	1000	5385	4986

Fig. 4. Hertzian contact pressure for 10 kN axial load + 10% radial force

Beside the significant change of the contact pressure (increase by 33%) the radial forces significantly change the ball contact kinematic, too. Consequence will be higher internal friction caused by 4-point contact and ball to ball contacts which reduces efficiency slightly and leads to tighter requirements for the lubrication.

3.1.2 Influence of ball nut and ball chain orientation relative to the load introduction vector

According to ISO3408 Part 5 [2], static and dynamic load ratings are calculated only from geometrical characteristics such as pitch circle and ball diameter, component accuracies like osculation, and the number of loaded balls. The orientation in relation to external axial and radial loads (see Fig. 5) as well as the position of the ball chains to each other (see Fig. 6) are not considered at all.

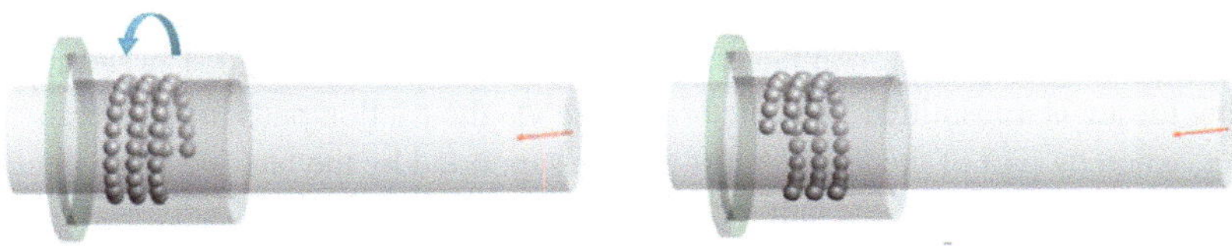

Fig. 5. Shift of ball start angle 0° to 360°

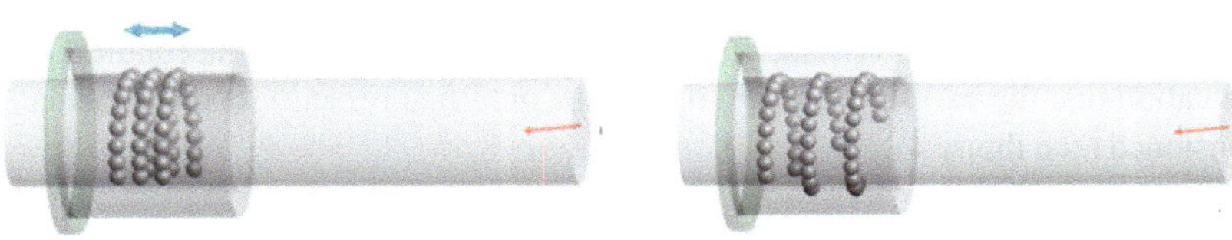

Fig. 6. Ball circuit position shift 0° to 360 °

Based on the calculation example for the Hertzian pressure safety factor limits (Fig. 7), in which both the ball start positions as well as the ball chain arrangement were varied fully factorially, it can be seen that the variance from min to max pressure can be up to 25%. Taking this insight into account, significant optimizations of the BSD in the overall system can be achieved.

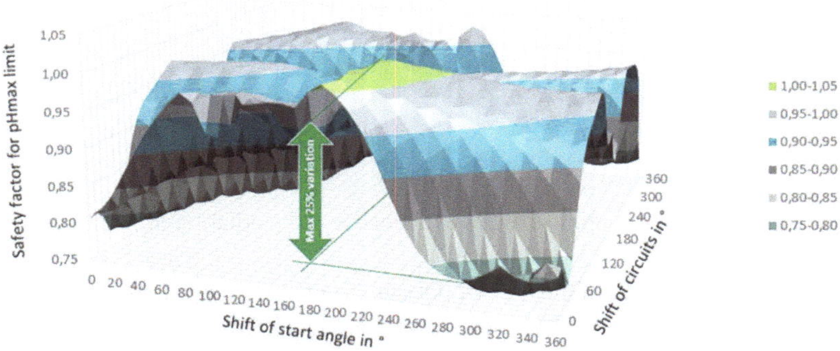

Fig. 7. Variation of hertzian pressure depending on start and ball chain position

3.1.3 Optimization loop internal ball nut design

If the principal ball nut concept (number of circuits, recirculation variant, number of balls etc.) is finalized and the first draft 3D model of the ball nut is created deeper investigations needs to be started. To get a suitable, robust and feasible design state of the art FEA tools are used.

Figure 8 shows the set of balls of a ball screw at the bore hole which is required for ball return. The balls passing on the raceway flank weakened by the bore cause an alternating bending stress, which is investigated by suitable FEM tools. Considering that the load collective derived from the specification, load and fatigue strength are compared and critical points on the nut contour are identified in order to reduce them by appropriate measures. Final evidence will be provided within the framework of the DVP during development.

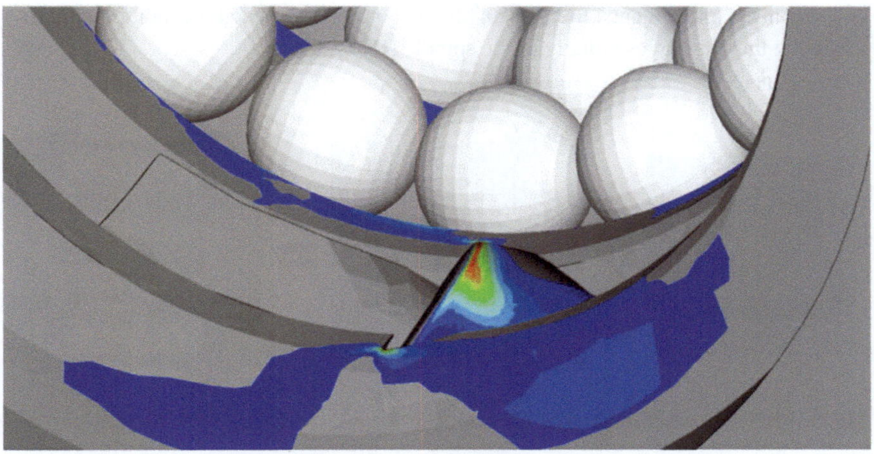

Fig. 8. Predictive analytic framework to optimize internal ball nut design

3.2 Function of Synchronized Planetary Roller Gear (SPRG) and its implementation in rear wheel steering

In the following the adaptions of the SPRG to rear wheel steering based on requirements are discussed. The mechanical efficiency in movement or standstill for a rear wheel steering is under 50 % due to the requirement to not move if the actor is in safe state or turned off.

To better understand the importance of and influences on efficiency, the function and especially movements inside the SPRG need to be discussed briefly. The SPRG is a special form of the PRG (see figure 9). In general, the PRG works like a planetary gear: the middle part of the planets (1) rolls on the spindle (2) while the outer parts of the planets roll on the nut (3). The spindle can move only transversal. The rotation of the spindle is blocked with an anti-rotational lock. When the nut is turned around the spindle, the planets, and the planetary carriers (4) rotate around the middle axis of the PRG. Neither the planets nor the nut but only the spindle have a pitch. Due to this pitch, the rotational movement is converted to a translational movement. To prevent a blocking, the planetary carriers hold the planets on equidistant distance to each other. A possible movement of the planetary carriers' inwards is blocked by the planets. The movement of the planetary carriers' outwards is limited by retaining rings (5) inside the nut. The PRG is highly statically overdetermined with several hundreds of con-

tacts. Between planets/nut and between planets/spindle are rolling contacts while between planets/planetary carriers are sliding contacts. Depending on load and design it is even possible that both sides of the grooves are simultaneously in contact.
In the PRG the nut is turned against the spindle and the carriers can move freely while in the SPRG the planetary carriers are coupled, turned against the spindle and the nut can rotate freely. For the rear wheel steering a belt is transmitting the movement of the motor to the pulley (7) of the synchronization. From the pulley this movement is transmitted over the flanges (6) to the planetary carriers. Additionally, there are several secondary movements: the planets tilt (A), yaw (B) and oscillate along their axis (C) while the nut moves radially to the spindle (D).

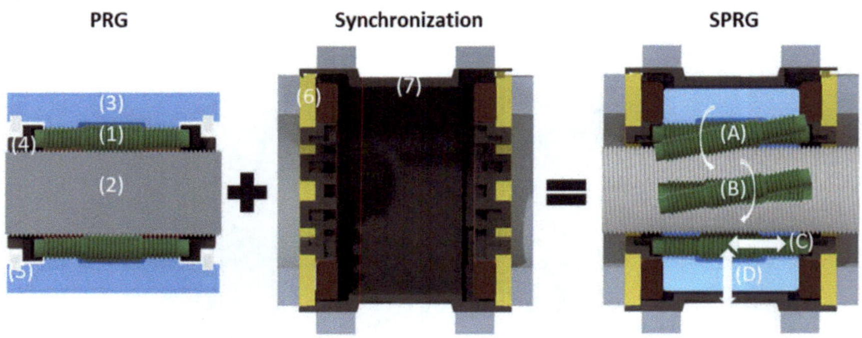

Fig. 9. PRG and synchronization form SPRG

Schaeffler has optimized the geometry for the highest power density – axial load capacity per installation space. Based on the optimum tooth width and the number of planets required to achieve the required load capacities it is possible to use either a single or a double-threaded spindle. A single-threaded SPRG in the needed size is self-locking while a double threaded is not. Regarding the installation space for the belt drive due to the needed transmission ratio the double-threaded spindle is far more advantageous. Additionally, the SPRG with single-threaded spindle needs double the turning rate and therefore is beyond the maximum reasonable speeds for good acoustics. The solution of a motor with double turning rate together with the double-threaded spindle has the same acoustic disadvantages. Schaeffler therefore choose the double-threaded non-self-locking spindle in combination with a reduction in system efficiency.

Figure 10 shows on the left possible areas to decrease system efficiency. (A) means the system efficiency would be reduced to achieve system self-locking by adapting the efficiency in the motor or the belt. To ensure that belt jumps do not also lead to a loss of self-locking, Schaeffler has decided not to achieve self-locking of the overall system by adapting the subsystems connected to (A). Furthermore, different belt ratios and motors are used in different applications. Adapting all these systems would be far more com-plex than adapting the identical core components of the SPRG.

Adaptations at positions (C) to (F) have a common challenge: If the generated braking torques rotate around the spindle axis, they are not supported by the bearing at (B). This means that these braking tor-ques around the spindle axis reduce the efficiency in the dynamics but do not improve the self-locking at standstill. Only the proportion of braking torques that explicitly do not rotate around the spindle axis improve the self-locking at standstill. It is therefore a potential to reduce the efficiency in the bearings at position (B) because this support problem does not exist there. An ideal combination of SPRG and bearings with slightly lowered efficiency allows high system efficiency near the 50 %, high power density and excellent NVH behavior without endangering the self-locking. Additionally rolling contacts instead of sliding contacts in the main flux of force reduce wear significantly compared to sliding only components like trapezoidal screw drives.

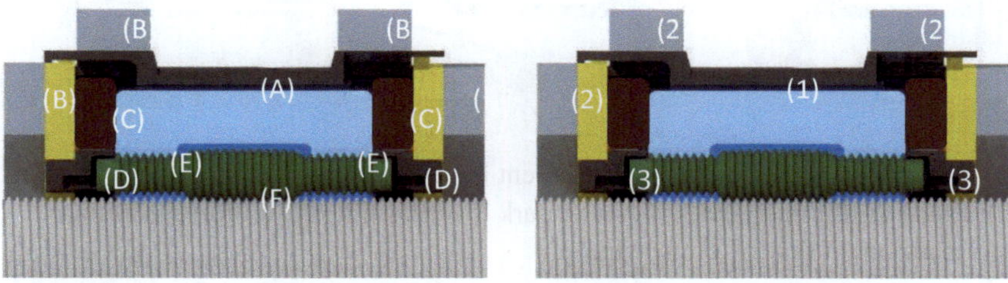

Fig. 10. Possible areas to decrease efficiency on the left and adapted components depending on customer requirements on the right

Depending on the customer requirements the SPRG is adapted further (see figure 10 on the right): First, the gear ratio of the belt is changed by adapting the number of teeth on the SPRG pulley. Second, de-pending on loads and needed stroke, the two axial and two radial roller bearings are replaced by two angular contact roller bearings. Third, for applications resulting in high wear the synthetic material of the planetary carriers is replaced by steel.

3.3 Influence of internal SPRG preload

A preload inside the SPRG is one of the possibilities to change the overall efficiency. Like bearings a x- or o-arrangement can be applied (see figure 11). To achieve a preload in x-arrangement the threads on the nut are moved inwards (yellow) and to achieve a preload in o-arrangement the threads on the nut are moved outwards (white). Additionally, there can be a virtual clearance. Due to tolerances in the production like bending of the planets the virtual clearance is normally replaced by a low stiffness at low loads – the components tolerances result in a small preload.
An external axial load higher than the preload will mainly be transmitted over one side of the nut (see dark blue line in figure 11). In most cases a smaller part (see orange line in figure 11) will be transmitted over the other side of the nut. Due to the long lever the side where the smaller part of the axial load is transmitted still has significant influences especially on the stability of the planet movement – e.g., if the

planet oscillates in tilting and yawing it can hit on this part of the nut on both side of the threads and therefore create high accelerations on the planet itself. Therefore, the positioning of the left and right side of the threads of the nut and with this the preload has significant influences on the behavior of the SPRG. For further explanations of the function, dynamics, and stabilities of (S)PRG it is referred to [3] and [4]. An example for the main flux of force is shown in the right side of figure 11. In an overdetermined system like the SPRG all main and many side forces significantly change the system behavior. For example a change in preload is influenced by and influences the load in the contacts of the planetary carriers – it can even result in the driving torque being only transmitted by one and not both planetary carriers to the planets.

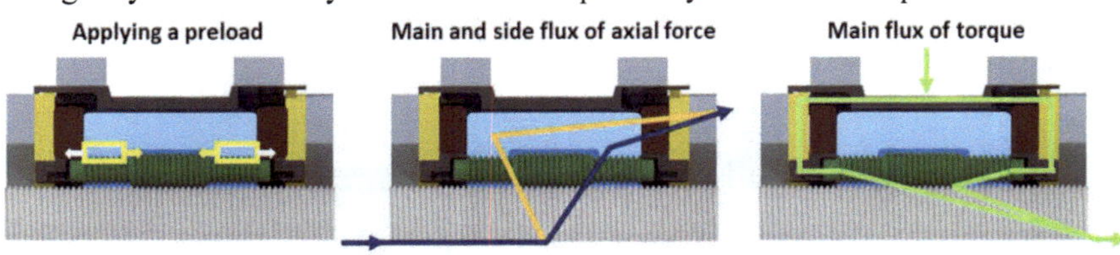

Fig. 11. Applying a preload in x-arrangement in yellow and in o-arrangement in white (left), examples for main flux of force in dark blue and side flux of force in orange (middle) and main flux of torque (right)

Each planet has slightly different production deviations. Therefore, the ideal position for each planet – the position in which the axial load on the SPRG is distributed even on all planets – is different for each planet. By individual and different tilting and yawing of the planets a near ideal force distribution be-tween the planets can be achieved. These ideal positions change under different load situations. For ex-ample, if the direction of the axial load on the SPRG is changed, the other sides of the threads of the components are loaded. Due to different production deviations on both sides of the threads, the ideal positions of the planets change depending on the direction of the load. In conclusion, a minimal movement space for the planets is needed.

With the help of multi-body simulations (MBS), effects of preload on the SPRG will be discussed below. For this purpose, MBS of the core of the SPRG with rigid bodies were set up with the boundary conditions shown in figure 12. The position and rotation of the planets is only defined by their contacts with the other components. The SPRG was loaded with e.g., 2 kN, and a forced rotational movement of the carriers was specified, which generates a feed rate of the spindle of 50 mm/s. The preload inside the nut was adapted by changing the geometry according to the previous description.

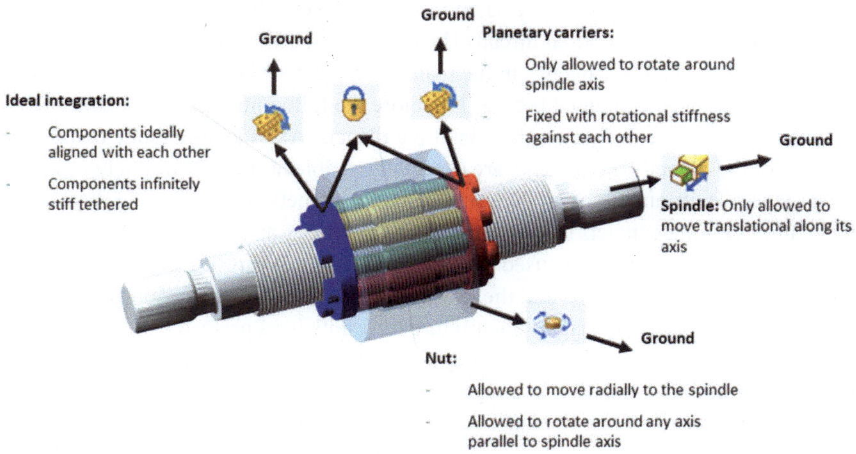

Fig. 12. Boundary conditions of the MBS

Figure 13 shows the efficiencies and tilting/ yawing angles of the planets over different preloads in x-arrangement. An increasing preload decreases the efficiency of the SPRG. A preload changes the basic friction and therefore has a great effect on the efficiency at lower axial forces. At low preloads the non-linearity of the efficiency is due to the change in the tilting of the planets and at higher preload the non-linearity is due to the change in yawing of the planets. In conclusion, it is shown that depending on their external and internal loads the planets have different stability position. Due to the clearance in the model the difference between the preload at 0 μm and at 60 μm is relatively small.

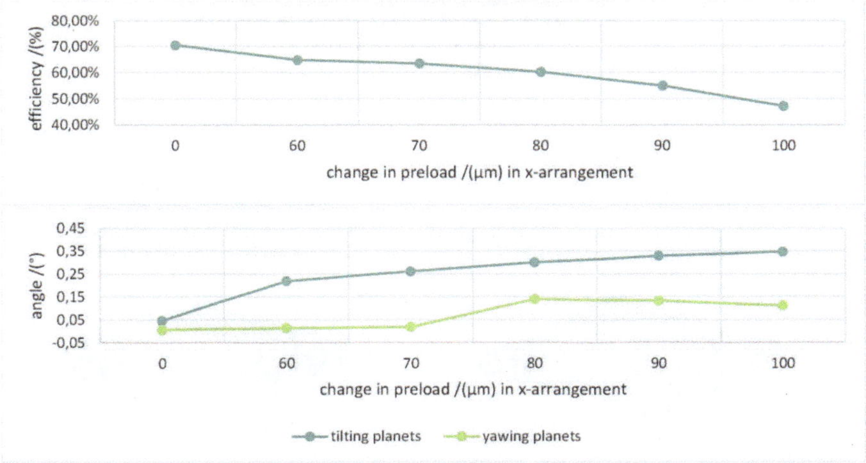

Fig. 13. Efficiency over preload (top) and tilting/ yawing angles of the planets over preload (down)

The overall SPRG behaviour is based on the movement of the planets and the load distributions on each individual thread. The decrease in efficiency by the preload is visible in these contacts. Changes in the planet's movements – their tilting and yawing has already been explained before. Therefore the following discussion concentrates on forces in the individual threads of the planets. Axial loads and the consequences of the planet's movements are especially visible at the contacts between the nut and the planets and between the spindle and the planets. The following discussion concentrates on the contacts between spindle and planet 1 as example. The results are similar for all the other planets used in this SPRG. The top part of figure 14 shows the naming scheme of the threads and the lower part the result for no pre-load on the left and 100 μm preload on the right. Each figure of results for a preload consists of two planes: first the upper part is a 2d column diagram showing the frequency of a certain normal force and its z-position in a regarding the load quasistatic time frame. The z-position is measured along the planet's rotation axis and is used to identify on which thread the contact between the two bodies happens. The lower part of each sub-figure of figure 14 is a heat map showing the same information.
Without a preload in the nut, the load distribution between the windings D2 to D11 is approximately line-ar. With the decrease of the in average transmitted load of each thread, the width of the frequency distribution of each thread is decreasing. In comparison with 100 μm preload, the load distribution of the windings follows an exponential function. The highly loaded windings are loaded even higher while the low loaded windings are loaded lower compared to no preload. Similar to [3] for the variation of the radial clearance for the PRG a reduction in lifetime is the consequence. Additionally at 100 μm preload contacts on both sides of the threads can be seen – e.g. C10 and C11.

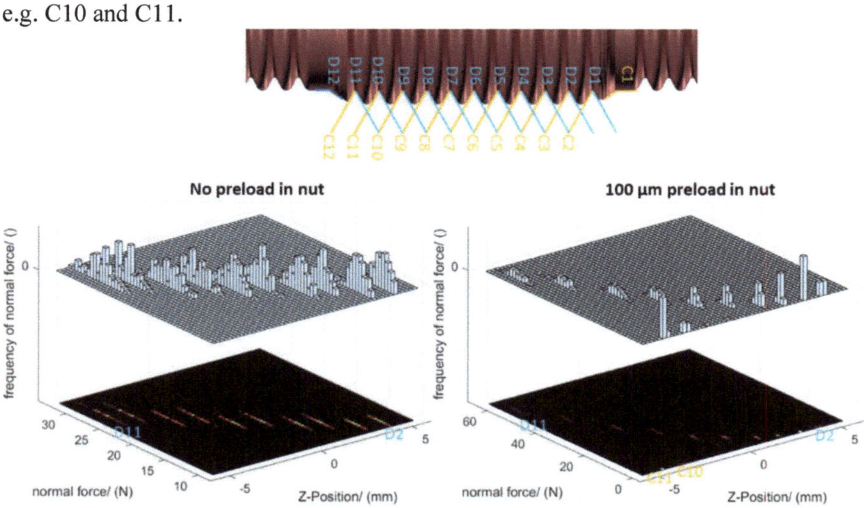

Fig. 14. Frequency of normal forces at contacts between spindle and planet 1 at quasistatic loads and no preload (left) / 100 μm preload (right)

4. Summary and outlook

The automotive industry has always been in a constant state of change. At present, the change is being pushed forward to a large extent by stricter climate targets which are driving tougher regulations, e.g. exhaust gas and particulate emissions. One effect these regulations have had on the industry, is stopping the development of pure combustion engines announced by suppliers and OEMs in favor of hybrid and purely battery-powered electric vehicles. As a result, numerous established applications had to be re-thought, as auxiliary media such as the vacuum were no longer available and belt-driven auxiliary devices were no longer classified as expedient. Regardless of future drive technology, these industry changes have strengthened the development of Rot-trans-converters in particular by chassis applications which are an integral part of the automotive industry.

In the field of x-by-wire applications, especially for steering and braking, developments have been underway for some time in which energy-saving "power on demand" actuators are being developed. The core elements for this are ECUs, electric motors and rot-trans-converters.

Figure 15 shows a steer by wire application in different positions and functions – e.g. how rot-trans-converters can be used for a steering actuator and a hand wheel actuator. In the steering actuator the spindle of the rot-trans-converter is moved translationally to turn the front wheels of a car. In the shown hand wheel actuator, the axial and lateral position of the steering wheel is changed by rot-trans converters.

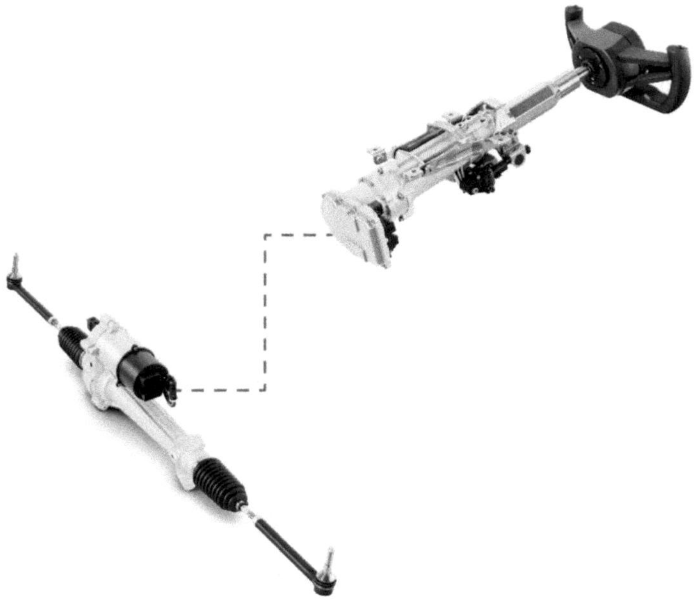

Fig 15. Steer by wire system

To show the adaptability of the SPRG an actuator for clutches shall be discussed. To open a clutch, a rather small translational movement at high forces is needed. Contrary for minimal needed design space an electric motor is rather using high turning speeds than high torques. Due to its high reduction ratio and power density, the SPRG fits perfectly to transform the rotation of the electric motor to the needed translational movement. When used to open a clutch, the SPRG will always work against springs closing the clutch. Therefore, there is always a minimal axial force as load on the SPRG. Additionally, this force is always in the same direction.
Like shown in figure 16, neither the main axial force nor the torque is transmitted over the left side of the nut. If the axial force goes always in the same direction, this will not change and therefore the left part of the nut, the left bearings, and the left part with threads of the planet are not needed. The system can be simplified like shown on the right of figure 5. This simplification to two instead of four bearings is only possible if the SPRG is driven without any lateral force. A hollow motor built around the SPRG is driving the SPRG for clutches instead of a belt of the shown general SPRG. Looking at the direction of the axial force, the planets of the general and the adapted SPRG for clutches will tilt: on their right side they will turn closer to the spindle and on their left side they will turn away from the spindle. To prevent the planets of moving out of the threads in the adapted SPRG, the left planetary carrier is needed to limit the radial movement of the planets. If the planet carrier on the right would be removed, the instability of the planet movement would increase greatly due to not enforcing the same distance between neighbouring planets on both sides. A high insta-

bility of the planet movement results in strongly fluctuating drive torques and decreased NVH behaviour. In extreme cases the planets move in a stick slip motion or block the component.

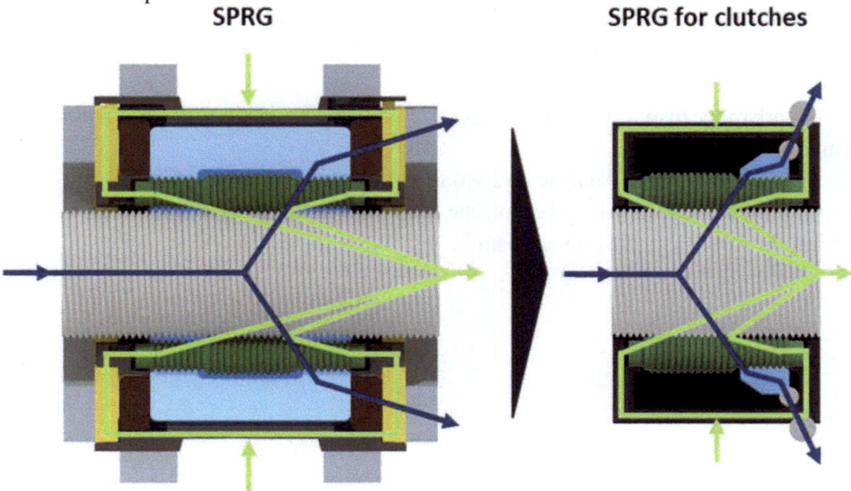

Fig. 16. Schematics of general SPRG and SPRG adapted for clutches. The shown adapted SPRG is realized in the hydrostatic modular clutch actuator by Schaeffler.

In conclusion, there are certainly numerous other possible applications to mention. One example could be a reinterpretation of an already established and series production system of adjusting ride height by means of an electromechanical actuator (see Fig. 17). Numerous advantages can be realized, such as load compensation, easier vehicle entry and exit, adjustment of ground clearance and above all, the reduction of CO_2 emissions by reducing the vehicle level when driving on the motorway.

Fig. 17. Electromechanical ride height adjustment utilizing a ball screw drive

References

[1] Pfeffer, P. und Harrer, M. (2013), Lenkungshandbuch Springer Verlag
[2] ISO 3408-5, Static and dynamic axial load ratings and operational life
[3] Merz, S. (2023), Analyse der Kinematik und Kinetik von Planetenwälzgewindetrieben. Dissertation,
KIT – Karlsruher Institut für Technologie, Karlsruhe, wbk Institut für Produktionstechnik
[4] Balázs, M. (1999), Analyse und Modellbildung einer Planeten-Wälz-Gewindespindel. Dissertation, Technische Universität München, München, DLR Insititut für Robotik und Systemdynamik.

Holistic Evaluation Methodology of the User Experience of Driver Assistance Systems in the Combination of Human-Machine-Interface and Functional Performance

Seda Aydogdu[1], Dr. Thomas Kersten[2] [0009-0005-8386-8925], Kevin Schuler[3], Prof. Bernhard Schick[3] and Gioele Micheli[1] [0009-0003-6747-1050]

[1] MdynamiX AG, Junkerstraße 4, 87734 Benningen, Germany
[2]Volkswagen do Brasil, Via Anchieta km 23,5, 09823-901 São Bernardo do Campo, Brasil
[3] Institut für Fahrerassistenz und automatisiertes Fahren, Junkerstraße 1A, 87734 Benningen, Germany
seda.aydogdu@mdynamix.de, bernhard.schick@hs-kempten.de, thomas.kersten@volkswagen.com.br

Abstract. Driver assistance systems aim to make driving safer and more comfortable while increasingly used in all vehicle classes and almost all markets. However, current research results continue to show low customer satisfaction and user acceptance. This is mainly due to the fact that the user experience and the holistic consideration of driver assistance functions as well as the human-machine interface (HMI) were neglected during development. A consequence of this is a low level of utilization, resulting in unused safety potential.

In this paper, we present a comprehensive evaluation method for driver assistance systems in terms of the combination of HMI and functional performance, especially for the Lane Keeping Assist (LKA) and Adaptive Cruise Control (ACC) systems. For this purpose, a test person study was conducted in close cooperation with Volkswagen Brazil and MdynamiX. The aim of this approach is to identify weak points of driver assistance functions in combination with display and operation (HMI). Based on these findings, concrete measures are derived to increase customer satisfaction, to increase acceptance during use and thus also to increase the degree of use.

In the context of this test person study, the complete user experience (before, during, after the journey) of drivers during a real journey with two different vehicles was surveyed by means of questionnaires. In addition, the functional performance of LKA and ACC was recorded by means of objective data collection. Thus, this work represents a holistic methodology for user-centered development.

The work highlights the need for advanced development of function, operation, display and warnings with a strong focus on the needs of customers to enable clear communication regarding system state and boundaries. In this way, functions can be developed that increase both the level of use and safety, as well as driving comfort.

P. E. Pfeffer (Ed.), *15th International Munich Chassis Symposium 2024*,
Proceedings, https://doi.org/10.1007/978-3-662-71064-7_6

First and second tier suppliers, but also service providers for "user experience", "usability" and "human-centered development", benefit from this work. The findings are crucial for optimizing driver assistance systems and better meeting customer expectations.

The study is characterized by the holistic view of HMI and function from the customer's perspective. Through this approach, targeted improvements can be made to meet customer requirements and expectations.

Keywords: ADAS, Human Factors, Adaptive Cruise Control, Lane Keeping Assistance System, HMI.

1 Introduction

ADAS functions are getting more and more important in the South American Car market. This is driven on the one hand by South American legislation like Latin NCAP or MoVer (Mobilidade Verde – Green Mobility) which is an energy efficiency law. On the other hand the customers can be convinced to use ADAS functions due to safety and comfort aspects. Taking over platform solutions (e.g. from Europe or Asia) to South America as many OEMs are doing and just doing some market application in a very cost sensitive market does not match the expectations of the majority of typical customers. The following approach leads to customer oriented and tailored ADAS solutions for the South American region and can be easily adapted to other regions with different cultures. Taking the Voice of the customer as a focal point and essential pillar into the development process and using objective values requires also a change in the mindset of development engineers and optimization of the development process from OEM point of view.

2 State of the art

2.1 SAM Region in numbers

Looking to some numbers for the SAM region, we are talking about 500 million people in 29 countries, Central America included. Regarding the amount of sold cars per year, the South American market sells about 4 million cars, which is about 5% of the world-wide car market. Taking the world-wide biggest car market, namely China as a reference for innovation and technology growth, the SAM is on a similar level: With 1.12 mobile connections [1] (China 1.18) per inhabitant, 80.6% Internet penetration [2] (China 73.7%) and 71.3% social media penetration [2] (China 72%). The observed increasingly high technological affinity might indicate a positive reception of well-working ADAS – equipped vehicles in the SAM market.

2.2 ADAS Development at SAM - Overview

In general, the South American Passenger Car market is mainly a market of small cars (A0 and A segment) due to the cost sensitivity if usual passenger cars are considered and not cars like e.g. Pickups. Various ADAS Systems from L0 to L2 have been established in the last years like ACC, AEB, LDW, LKA, Travel assist, RTCA or Park Assist. Hence for example, the region is gathering for camera based instead of Radar-based systems for the functions mentioned above. Looking to the future, the task for development engineers is to climb the mountain of L2+ over even L2++ systems which are cloud-based and usually very expensive under highly cost restrictive circumstances.

2.3 Drivers and traffic conditions at SAM

Before talking about the customers' needs, we have to consider the environmental boundary conditions which are very particular in South America. Among them there are bad road conditions, hard traffic, long distances, high rate of speed bumpers speed control, and high altitudes. Travelling long distances by car is a main driver for ACC, ACC stop and go, Travel assists and LKAS systems. Taking part of huge traffic jams in megacities like Sao Paulo with 20 million people such systems like ACC, stop and go, FCW, and other collision warning systems are warmly welcome due to crossing motorbikes at high speed. Summarizing this situation, the task to the OEM is to generate customer benefit in terms of safety and comfort while taking cultural aspects into account.

Finally, the legislation like Latin NCAP and MoVer in Brazil which generates incentives to the OEMS in case of equipping the cars with ADAS functions are helping to penetrate the South American Automotive industry with ADAS systems.

2.4 Development Process

Nevertheless, assuming to have the best ADAS system in terms of functions won't be accepted in case the usage is bad and complicated and the HMI has no clear communication. Therefore, it is essential to create a new development process which takes care of all aspects regarding functionality, usability, HMI communication and cultural backgrounds. Furthermore, it is of utmost interest to relate the subject evaluation to objective KPIs. Based on that it is planned to an objective measurement of KPIs identify the correlation to the subjective impressions. This task will be done in the next month in a further project.

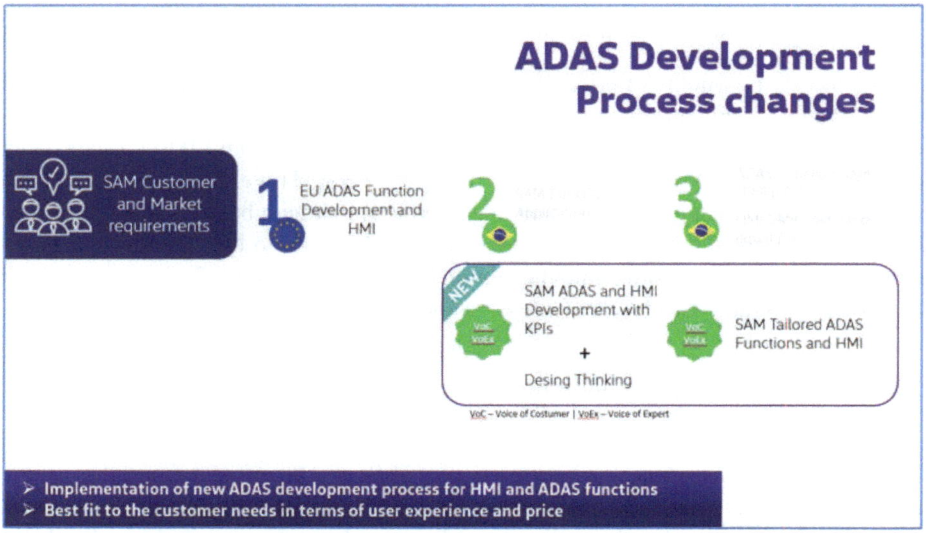

Fig. 1. Improvement of ADAS development Process at Volkswagen do Brasil

2.5 Research questions

Aim of the research is creating a methodology to develop vehicles holistically (HMI's UX and usability, functional performance) and testing whether the approach yields better results in improving driver's satisfaction. This led us to consider the following questions:

RQ1: Can a methodology be developed to compare vehicles holistically considering both Functions and HMI?

RQ2: Does the benchmarking vehicle perform better or worse compared to the Taos in with respect to ADAS and HMI?

RQ3: Which optimization tasks can be derived from participants?

3 Methods

3.1 Experiment design

In this experiment, the ADAS (LKA, ACC) of a VW Taos (model year: 2024) and of a benchmarking vehicle (model year: 2024) are compared in terms of functional and of HMI performance with 43 (scheduled: 50) participants. The benchmarking vehicle was selected based on similarity of class and segment, of availability of relevant ADAS, and of being perceived as a vehicle with capabilities to learn from. Of the 43 participants, five were considered as experts. Results recorded include both quantitative subjective questionnaires as well as qualitative, unguided interviews. Observed ADAS functions include a LKA and an ACC.

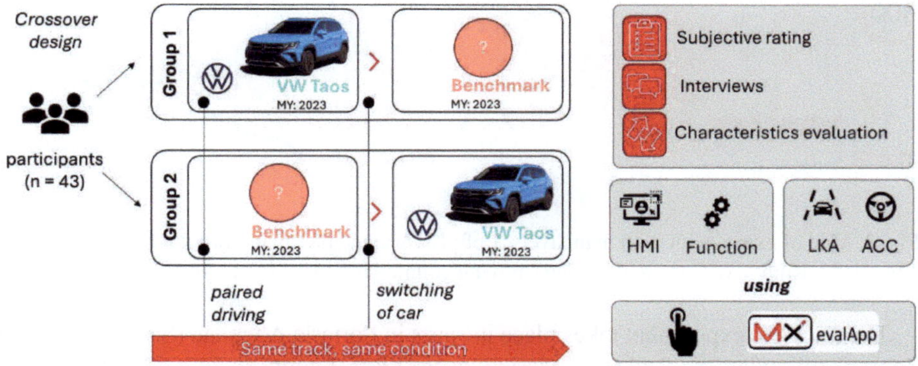

Fig. 2. Overview on the experiment design.

The trial is staged as a cross-over (AB / BA) design (compare with Fig. 2) [3]. The first group (AB) tested the VW Taos first (Period 1) and then the benchmarking vehicle (Period 2), the other group (BA) vice versa. Both participants start driving their assigned cars simultaneously to reduce between-subject variance due to varying traffic conditions. After completing the test sequence, a washout period of 30 minutes is scheduled. After the washout period, participants switch seats and perform the second round of tests with the other vehicle respectively.

Participants were sampled using a friends-and-family approach. A public recruiting was avoided for safety concerns. Participants were not actively balanced for age or gender, although both demographics were recorded. Additionally, participant's employment occupation and education levels were requested, to match participants with customer target group. Participants' prior knowledge was recorded with a pre-evaluation questionnaire. Among the information recorded, vehicle ownership, driving experience in km/year and years, ownership of ADAS, and usage of ADAS systems was recorded.

For each vehicle, there is one experiment lead respectively. The task of the experiment lead is to navigate the participating driver, instruct to initiate maneuvers, record results, and conduct the interview.

3.2 Track and maneuvers

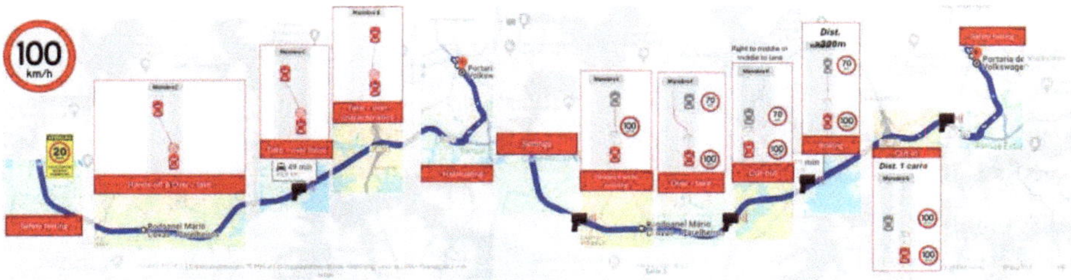

Fig. 3. Map of experiment with maneuvers. Left: LKA maneuver map (moving clockwise / right – to – left). Right: ACC maneuver map (moving counterclockwise / left – to – right).

The track the experiment takes place in starts in Portaria Anchieta (São Bernardo do Campo) following the Rodoanel Mário Covas (SP-021) until in Exit 23 (LKA, Fig 3 left), where the route follows back to the beginning (ACC, Fig. 3 right). It was selected considering the boundary conditions: Safety concerns, route with low probability of traffic jam, optimal length for number of maneuvers, minimum amount of tall collection stations, allowed speed required for ADAS system activation. Experiment leads described weather (subjectively) as *Bad* in 8% (n = 4) of cases, and in 12% (n = 6) of cases traffic was subjectively described as *Heavy*. In total, five (four) maneuvers were considered for evaluating the ACC (LKA). Figure 4 gives an overview on maneuvers considered. A written explanation can be found in Tables 2 and 3.

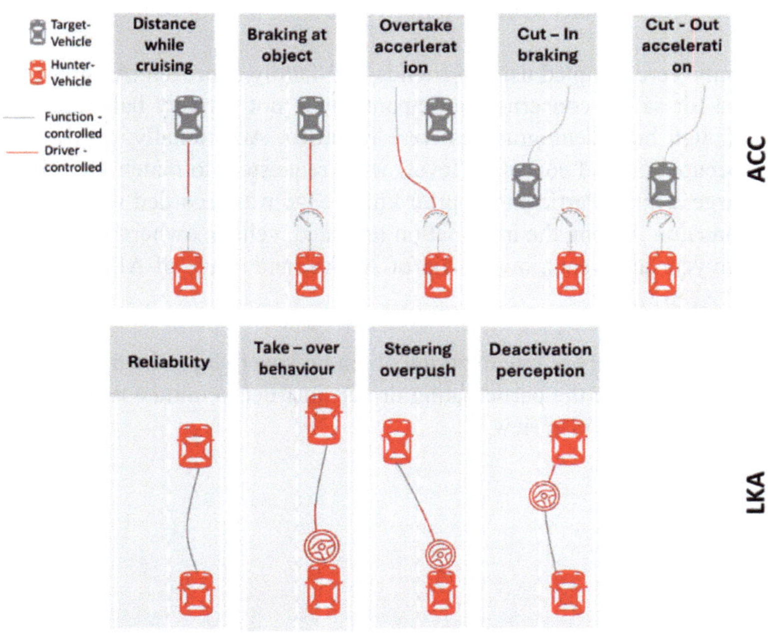

Fig. 4. Overview of maneuvers considered for evaluation.

3.3 Measurements and Criteria

Two types of data were recorded in this experiment using the MXeval App: Qualitative and quantitative data. Quantitative data includes demographic data, subjective ratings of both function and HMI, and evaluations of characteristic information. All subjective are based on a 1 – item 10 – point Likert scales. Additionally, for certain criteria based on maneuvers, a subjective characterization based on the following scales is used:

- **Timing**: How well the ADAS maneuver was timed, rated from very early to very late.
- **Strength**: How well the acceleration force was perceived, rated subjectively from very weak to very early.
- **Distance:** How well was the distance perceived that the vehicle kept from an approaching (also referred to as *target*) vehicle in a maneuver, rated subjectively from very far to very close.

These dimensions can describe parameters of a maneuver that allow creating tradeoff charts. Interviews are the only type of qualitative data recorded. Data was recorded in three phases: Before test drive (Demographic and Pre-Questionnaire), after maneuver (ACC & LKA criteria), and after experiment (HMI, LKA reliability). Criteria are derived from [4] [5] [6].

HMI criteria

The HMI serves as interface with the automated car. Criteria in this section are mainly limited to functions pertaining to the ADAS functions, therefore mainly belonging to the aHMI (automation HMI, according to the taxonomy in [7]). Table 1 provides an overview of the criteria used to rate both the LKA's and the ACC's HMI.

Table 1. Overview of HMI criteria used. (AT = After test drive, LS = Likert scale)

Criterion	Explanation	Measured
Activation usability (only ACC)	Rating general impression and usability of activation user interface	AT, LS
Adjustment usability (only ACC)	Rating general impression and usability of adjustment user interface	AT, LS
Perceived safety	Subjectively perceived degree of safety when system is active	AT, LS
Display	Subjective rating of usability and attractiveness of the display concept	AT, LS
Supervision workload	Describes workload required to supervise (monitor) the system	AT, LS
Warning system	Describes design of acoustic, optical, and haptic warnings of the ADAS	AT, LS
Ability to self-explain	Describes system usability without explanations	AT, LS
Quality to relieve	Describes subjective degree of relief from the active system	AT, LS

Single Ease Questionnaire (SEQ) [8]	Rating of overall usability in one question	AT, LS
System Usability Score (SUS) [9]	Ten-item question score to assess general usability	AT, LS

Functional criteria ACC

The ACC is an ADAS that allows "following a forward vehicle ahead at an appropriate distance" (ISO 15622 [10]). For evaluating the ACC, only criteria based on maneuvers performed are considered. Table 2 outlines the maneuvers evaluated for ACC.

Table 2 Overview of ACC criteria used. (AM = After maneuver, LS = Likert scale, FT = Force timing characteristics, D = Distance characteristics)

Criterion	Explanation	Measured
Breaking at object	Strength and intensity of breaking to avoid collision with approaching vehicle	AM, LS + FT
Cut – out acceleration	Strength of acceleration when target vehicle cut-out with ACC active	AM, LS + FT
Cut – In Braking	Timing and acceleration of braking to avoid collision with cutting in target vehicle	AM, LS + FT
Overtake acceleration	Timing and acceleration of the noticeable acceleration with ACC active	AM, LS + FT
Cruising distance	Describes Distance between target and hunter that can be set while driving	AM, LS + D

Functional Criteria LKAS

The LKA (cf. ISO 11270 [11]) is a different ADAS that can control the vehicle longitudinally and laterally to ensure it stays within the lane boundaries. Table 3 gives an overview of criteria used to rate the LKA.

Table 3 Overview of LKA criteria used. (AM = After maneuver, AT = After test drive, LS = Likert scale, FT = Force timing characteristics, D = Distance characteristics)

Criterion	Explanation	Measured
Take – Over characteristics	Criterion describes system intervention when approaching and recentering from lane limits	AM, LS + FT
Perception of Warnings	Describes how well the warnings were perceived, or if they were perceived at all	AM, LS + D

Steering take-over force	Describes steering torque required to overdrive lane markings with LKA active without indicators	AM, LS
Deactivation perceptions	System capability of reaching over control when leaving its ODD so driver can safely control vehicle	AM, LS
Warning sequence while deactivating	Describes how warnings were perceived optically, acoustically, and haptically with respect to the situationally adapted salience	AM, LS
Perceived reliability	Criterion describes consistent and precise ability to hold in lane limits without wrong warnings in all conditions	AT, LS

3.4 Hypothesis and Evaluation

As the aim of the study is comparing the vehicles as a benchmark, all criteria's values are compared in non-directional hypothesis. Accordingly, the following set of hypotheses is defined, where Criterion C_i denotes all Criteria listed in tables one, two, and three, leading to **29 pairwise criterion comparisons** (8 criteria (6) for HMI ACC (LKA), 5 criteria (6) for functions ACC (LKA)).

$H_{0,i}$: The VW Taos' average subjective score of Criterion C_i doesn't differ from the corresponding score of the benchmarking vehicle.

The (AB/BA, two-period, two-sequence) cross-over design is evaluated using the approach described in [3]. Accordingly, tests are performed to determine whether a carryover effect exists by comparing within-subject-sums. A carryover effect is assumed below a significance level of $\alpha = 0.10$, according to literature [3]. In case no carryover effect could be verified, a test for treatment effects is conducted by comparing within-subject differences using a significance level of $\alpha = 0.05$. All evaluation is performed using the R programming language to ensure a reliable, reproducible, and transparent evaluation procedure. Interviews are summarized using a LLM approach of synthesis.

4 Results

4.1 Limitations

As the procedure was implemented for the first time, some effects might have influenced the results. Mainly, the family and friends recruiting approach might have led to a distortion in which participants favorized the VW Taos over the benchmarking vehicle (Demand characteristics [12]). Similarly, the experiment leads' expectations being company internal might have further influenced the result favorizing the VW Taos.

The benchmarking vehicle performed worse or equal in nearly all (28/29) categories. Such a strong result can indicate the benchmarking process might have not considered the best system. Therefore, while results obtained from the study might reflect accurate comparisons, they don't help in understanding areas of improvement comparing with the best-in-category systems.

Furthermore, because the experiment was performed under real-life conditions, certain influences can't be controlled. In some instances, an experiment was postponed or split due to bad weather conditions. This led some participants to drive unpaired, causing the overall traffic conditions in both groups not to be symmetric. While influencing the results, issues like these seem to be the prize for ensuring real-life conditions.

4.2 Demographics and conditions

Participants' average age was 41.2 (SD = 12.9) years old; the gender ratio of participants was 48:52 female-to-male. Seven initially scheduled participants dropped out. Regarding the MDSI self-described themselves as average with a tendency to Calm – Defensive.

4.3 ADAS HMI

From criteria used to assess the HMI, the three subjectively most important categories, on average for the ACC were the perceived safety, the display, and the activation usability. For the LKA, the most important categories were the perceived safety, the display, and the supervision workload.

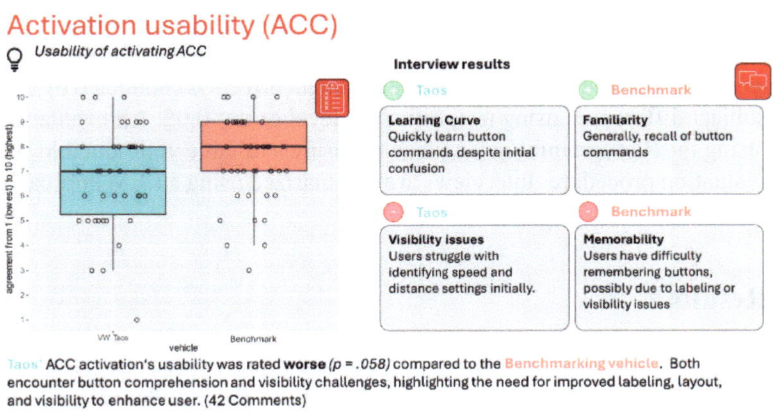

Fig. 5. Dashboard created for HMI – evaluation type criteria including an analysis on strengths and weaknesses.

To gain an overview on the results, an overview dashboard (cf. Fig. 5) is created were results of all significance testing of all categories are summarized. For each category, a detailed result dashboard is created. In it, a boxplot with all the responses overlayed is portrayed together with a Strength – and – Weaknesses synthesis of all comments to the respective field. A summary below helps management take decisions backed with hypothesis – tested results. This procedure allows quickly identifying fields of improvement and giving feedback on the spot. It was noted that some responses lack insight because interviewing questions were open.

4.4 ADAS functions

For the ACC criteria, participants preferred by the customers in order: Braking at object, Cut-In Braking, Distance while cruising, whereas for the case of the LKA the criteria preferred in order are reliability, steering intervention, and effort against intervention.

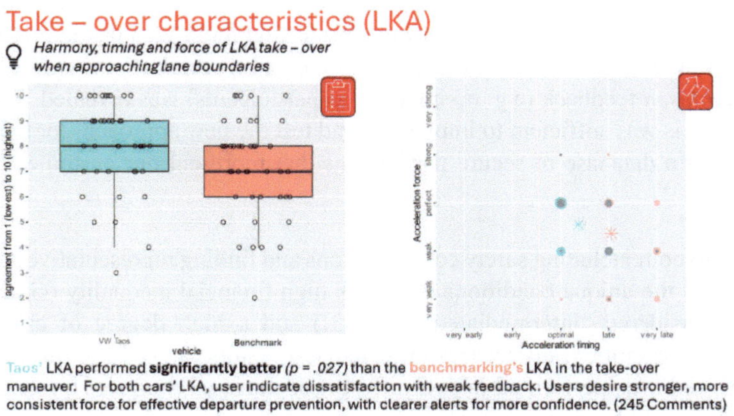

Fig. 6. Dashboard created for Function – evaluation type criteria including trade off charts. Bubbles represent frequency of characteristics combined, while stars represent a weighted average of responses.

The procedure for evaluation and improvement is the same in the case of functions like for the HMI. Like for the HMI, the in – detail dashboard (cf. Fig. 6) contains a boxplot as well as an actionable summary. The difference in this case is that for criteria in which characteristics are used (e.g. very early/very late), a trade off chart is displayed. They provide meaningful feedback to calibrate function behavior and allow tailoring functions to local market needs. An average deviation from the center (optimum) directly indicates a potential for improvement.

5 Discussion

The methodology developed in this work introduces a procedure to a part of development where none existed before. The procedure allows introducing a test procedure to ADAS (LKA and ACC) that raises both to its intentions of holistic development and being tailored to understanding the needs of customers in the region of SAM. This evaluation approach is vital because customers experience the vehicle as an entire system and not a part.

So, how did the method perform? According to us, the granularity in which criteria show potential for improvement is fine enough to identify concrete measures while not falling prone to losing the bigger picture. However, the largely one-sided test results revealed the need to implement a proper benchmarking approach to help select an optimal comparison vehicle to gain more feedback on potential for improvement.

The one-sided results also raised the question whether the friends-and family approach skewed participants opinions towards preferring the VW. For the recruiting, a conflict between the goals *safety* (avoidance of theft), *representativeness* of customer ship, and *unbiased* feedback (e.g. no internals as participants) was revealed. While the procedure as it is was sufficient to implement and test the new approach, for the future, the creation of a database of secure participants that represent our customer is being discussed.

The need to both including safety considerations and finding representative customer ship stem from the unique conditions in Brazil: high financial inequality (Gini coefficient 49.7 considered "intermediate-high" [13]) and a high degree of criminality. Whether and how cultural differences influence test results can only be answered if a test like this one was performed in two different countries, yet under similar circumstances. This would be a scenario where driving simulators would be of use. We also learned that it is important including local, mother tongue speaking experimenters for conducting tests to avoid communication problems and better predict participants' behavior.

For future projects, implementing a similar procedure in the format of an expert evaluation is planned. The expert evaluation would differ from the test person study by letting experts rate categories in more detail and objectively measuring KPIs. Especially for aspects like the HMI, this study's result results lacks indicating concrete ways of improvement. In this way, both designs can be used complimentary: One to find potential for improvement, and the other to understand what to change. In case it remains difficult to understand how these changes can be implemented, a design thinking phase could be introduced to identify solutions for remaining weak points. As an additional tool, certain safety-critical maneuvers can be tested in a driving simulator to collect information in an environment with less pressure. Once finished, such a holistic evaluation approach can be applied also for other and coming ADAS developments (e.g. parking assist) and other testing in which humans interact with the vehicle.

References

1. We Are Social. Special Report Digital 2023. Link: https://wearesocial.com/cn/blog/2023/01/digital-2023/ (Accessed 06.05.2024)
2. Data Reportal. Digital 2023 April Global Statshot Report. https://datareportal.com/reports/digital-2023-april-global-statshot (Accessed 06.05.2024)
3. Wellek S., Blettner M.: On the proper use of the crossover design in clinical trials: part 18 of a series on evaluation of scientific publications. Dtsch Arztebl Int. 2012 Apr;109(15):276-81. doi: 10.3238/arztebl.2012.0276. (2012)
4. Aydogdu S., Seidler C., and Schick B. Trust Is Good, Control Is Better? – The Influence of Head-Up Display on Customer Experience of Automated Lateral Vehicle Control. In HCI in Mobility, Transport, and Automotive Systems: First International Conference, MobiTAS 2019, Held as Part of the 21st HCI International Conference, HCII 2019, Orlando, FL, USA, July 26-31, 2019, Proceedings. Springer-Verlag, Berlin, Heidelberg, 190–207. https://doi.org/10.1007/978-3-030-22666-4_14 (2019)
5. Aydogdu S., Schick B., and Wolf M., “Claim and Reality? Lane keeping assistant - The Conflict between Expectation and Customer Experience,” in Fahrzeug- und Motorentechnik, Aachen. (2018)
6. Nesensohn, J., Lefèvre, S., Allgeier, D., Schick, B., Fuhr, F.. An Efficient Evaluation Method for Longitudinal Driver Assistance Systems within a Consistent KPI based Development Process. In: Pfeffer, P.E. (eds) 11th International Munich Chassis Symposium 2020. Proceedings. Springer Vieweg, Berlin, Heidelberg. https://doi.org/10.1007/978-3-662-63193-5_7. (2021)
7. Bengler K., Rettenmaier M., Fritz N., and Feierle A. From HMI to HMIs: Towards an HMI Framework for Automated Driving. Information 11, no. 2: 61. https://doi.org/10.3390/info11020061. (2020)
8. Sauro J. and Dumas J. Comparison of three one-question, post-task usability questionnaires. In Proceedings of the SIGCHI Conference on Human Factors in Computing Systems (CHI '09). Association for Computing Machinery, New York, NY, USA, 1599–1608. https://doi.org/10.1145/1518701.1518946 (2009)
9. Brooke, J. SUS: A Quick and Dirty Usability Scale. In: P.W. Jordan, B. Thomas, B.A. Weerdmeester & I.L. McClelland (Eds.), Usability Evaluation in Industry. London: Taylor & Francis, 189-194. (1996)
10. International Organization for Standardization. Intelligent transport systems: Adaptive cruise control systems – Performance requirements and test procedures. ISO Standard No. 15622:2018. (2018)
11. International Organization for Standardization. Intelligent transport systems: Lane keeping assistance systems (LKAS) – Performance requirements and test procedures. ISO Standard No. 11270:2014. (2014)
12. Orne, Martin T. Demand Characteristics and the concept of Quasi-Controls. in Artifacts in Behavioral Research: Robert Rosenthal and Ralph L. Rosnow's classic books : a re-issue of Artifact in behavioral research, Experimenter effects in behavioral research and The volunteer subject, pages 110 pp. (2009)
13. UNU WIDER. World Inequality Database – WIID. Link: https://www.wider.unu.edu/database/world-income-inequality-database-wiid#WIIDcomp (Accessed 06.05.2024)

Torque Vectoring Control for More Driving Fun

Zoltan Hankovszki, PhD[1], Jan Loof, PhD[2] and DI Michael Lipp[3]

[1,2,3] AVL List GmbH, Hans-List-Platz 1, 8020 Graz, Austria
[1]zoltan.hankovszki@avl.com
[2]jan.loof@avl.com
[3]michael.lipp@avl.com

Abstract. Multi-motor electric powertrains offer the possibility to set up the ideal drive torque distribution at the wheel in every situation. However, working out the control algorithm for this purpose is a major challenge. And this is only the first step if the final goal is to implement this control into a series-production passenger car. AVL's Torque Vectoring (TV) function family realizes a state-of-the-art vehicle motion control concept, which ensures good driving experience, but it also has to deal with classical validation steps, for example handling sudden friction value changes and avoiding unneeded stability losses, at the same time. Validated vehicle models in the AVL-VSM simulation environment shorten the development time.

Keywords: Torque Vectoring, Yaw-Rate-Controller, Yaw-Moment Controller

1 DRIVING FUN – STATE-OF-THE-ART TORQUE VECTORING

The developed TV algorithm consists of five main parts:

- vehicle state observers which continuously provide input data for the control elements
- a target vehicle model which generates a target yaw-rate
- a TV yaw-rate controller to control the lateral vehicle behavior
- a wheel-individual slip controller to utilize the grip potential
- a yaw-moment controller to combine the yaw-moment request with wheel torque limits by the wheel-slip controller and motors.

An overview of the TV architecture is shown in Fig. 1.

P. E. Pfeffer (Ed.), *15th International Munich Chassis Symposium 2024*,
Proceedings, https://doi.org/10.1007/978-3-662-71064-7_7

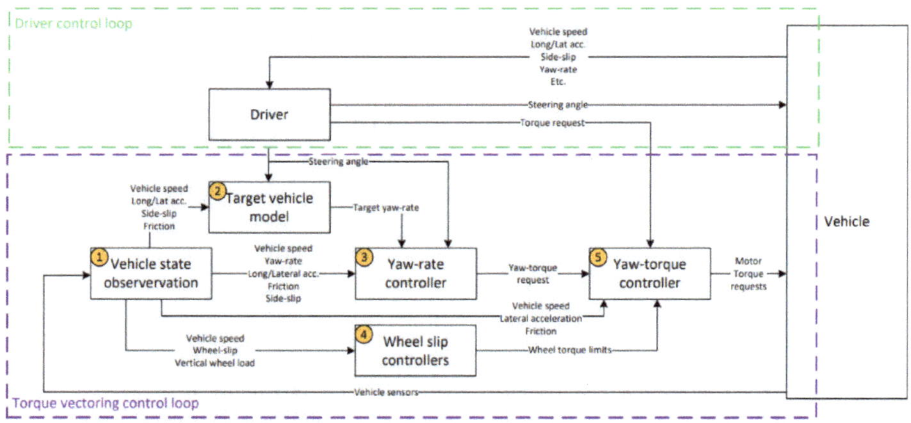

Fig. 1. Overview of the torque vectoring architecture with driver control loop and torque vectoring control loop (© AVL List GmbH)

There, two control loops can be distinguished. First, the driver controls the lateral vehicle behavior with the steering angle, making decisions based on vehicle states such as vehicle speed, longitudinal and lateral acceleration, side-slip angle of the vehicle, yaw rate, etc. Additionally, the driver applies a torque request via the acceleration pedal to control the speed of the vehicle. The second one is the TV control loop which uses a vehicle state estimation and several observers to gather quantities such as vehicle speed, longitudinal and lateral acceleration, road friction value, side-slip angle, wheel loads and wheel slip. A target vehicle model calculates a target yaw rate $r_{z,ref}$. which is based on the vehicle condition and the steering angle. It can be calibrated per drive mode to allow a more sporty or more comfortable driving behavior. In order to ensure that the target yaw rate is also feasible for the vehicle, a saturation limit is used, which is based on the friction value observer and the side-slip angle observer β, according to Lenzo [1]. Whenever the side-slip angles are small, the handling target yaw rate $r_{z,ref,h}$ is used. Whenever the side-slip angles get too large, a correction is made using the side-slip angle and the lateral acceleration to stabilize the vehicle, resulting in the stabilizing target yaw rate $r_{z,ref,s}$ according to Eq. 1:

$$r_{z,ref} = r_{z,ref,h} - F(\beta)\left(r_{z,ref,h} - r_{z,ref,s}\right) \tag{1}$$

Here, F as the arbitration factor is calculated according to Eq. 2, which depends on the side-slip angle:

$$F = \begin{cases} 0 & if\ |\beta| < \beta_{act} \\ \frac{k_1(|\beta| - \beta_{act})}{\beta_{th} - \beta_{act}} & if\ \beta_{act} \leq |\beta| \leq \beta_{th} \\ k_2 & otherwise \end{cases} \tag{2}$$

The yaw-rate controller consists of a pre-controlling (feedforward controller) which directly translates the steering wheel angle of the driver and the vehicle state into a yaw-moment request, Fig. 2.

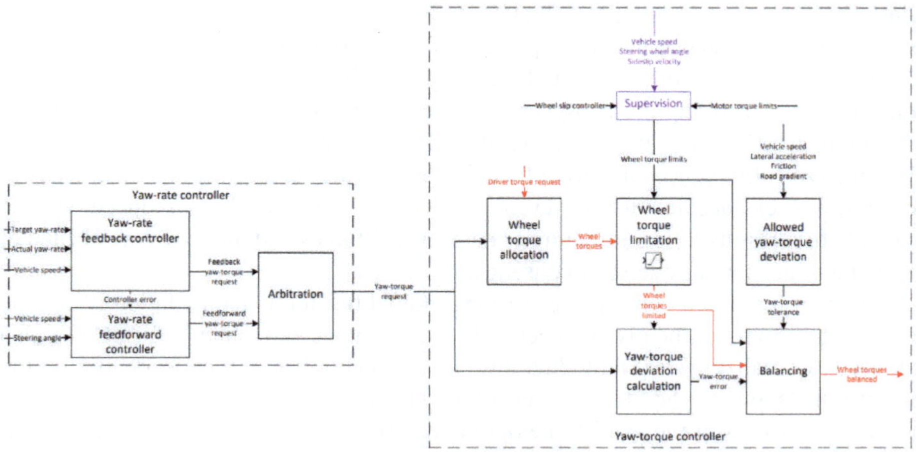

Fig. 2. Architecture of the yaw rate controller (left) and the yaw moment controller (right) – the red path shows the torque path (© AVL List GmbH)

An adaptive feedback controller evaluates the actual vehicle yaw-rate and makes additional corrections when this actual yaw-rate deviates from the target yaw-rate. The feedback and feedforward yaw moment are arbitrated to a yaw-moment request that is forwarded to the yaw-moment controller. The goal of the yaw-moment controller is to combine the driver torque request for the respective axle with the yaw-moment request while at the same point respecting the torque limits by the wheel-slip controller and the electric motors. To do this, the wheel torques are allocated by Eq. 3:

$$T_{w,i} = \frac{T_d}{2} \pm \frac{T_{yaw,r}}{\frac{1}{2}t_r} R_i \; with \; i \in \{l, r\} \tag{3}$$

Here, $T_{w,i}$ is the torque request of wheel i, where i can be left (l) or right (r), and T_d the entire torque of the axle. $T_{yaw,r}$ is the yaw-moment request, t_r is the track width of the axle and R_i is the tire radius of the respective wheel. Depending on the torque limits, the wheel torques are upper and lower limited, resulting in $T_{w,i,lim}$. These limited wheel torques are used to calculate an error yaw moment $T_{yaw,e}$ with Eq. 4:

$$T_{yaw,e} = T_{yaw,r} - \underbrace{\frac{1}{2}\left(\frac{T_{w,r,lim}}{R_r} - \frac{T_{w,r,lim}}{R_r}\right) t_r}_{\text{Applied yaw-torque}} \tag{4}$$

2 SUPERVISED DRIVING FUN

Since the TV system manipulates the longitudinal tire forces to generate yaw moment, it is worth to crosscheck the consequences of longitudinal tire force excitation [2, 3]. Electronic vehicle motion control (Electronic Stability Control, ESC) already has a long history controlling the subsystems brake and powertrain in order to generate a yaw moment. However, there is a significant difference between ESC and TV: The first one typically reduces the drive forces and/or creates brake forces, which leads to lowered vehicle acceleration or even deceleration. The resulting reduced vehicle speed in turn results in lower required lateral wheel forces and increases the chance of reducing understeer or oversteer consequently.

A further important consequence of longitudinal tire force manipulation is that the cornering stiffness of the given wheel is directly influenced. Unlike TV, ESC can also be used to realize wheelindividual tire force manipulation independently of the powertrain layout. In addition to generating the yaw moment, this also allows the cornering stiffness of the axles to be optimized. These two consequences ensure such a robustness, that relatively simple control strategies (for example starting an intervention only after yaw rate or vehicle side slip is reaching a limit) can successfully control understeer or oversteer situations. Under certain circumstances, a similar control strategy for a TV algorithm may not recognize a stability loss tendency early enough.

In contrast to ESC, TV control may happen continuously in a way that it is not disturbing for the driver. The effects of ESC and TV interventions are described in detail in Tables 1 and 2.

Table 1. Effects of ESC interventions (© AVL List GmbH)

Feature	**Effects of wheel individual braking (ESC)**		
	Rotation around yaw axis	**Longitudinal translation**	**Side-slip angle rotation**
Direct, time invariant	Manipulated yaw acceleration:	Reduced acceleration or even deceleration:	Manipulated side-slip angle velocity:
	+ Primary effect	– Side effect leads to axle load change in oversteer direction	+ Cornering stiffness balance change, selectable for understeer or oversteer
Derived, time variant	Yaw rate with cumulated manipulation:	Reduced velocity:	Side slip with cumulated manipulation:
	+ Temporary and progressive, feedback interventions	+ Lower velocity → lower side forces needed → improved tire reserves	+ Temporary and progressive, feedback interventions

Table 2. Effects of TV interventions (© AVL List GmbH)

Feature	Effects of rear axle lateral drive torque distribution (TV)		
	Rotation around yaw axis	**Longitudinal translation**	**Side-slip angle rotation**
Direct, time invariant	Manipulated yaw acceleration:	Acceleration:	Side-slip angle velocity tending more oversteer:
	+ Primary effect	+ Can be kept as the driver demands	– Cornering stiffness balance change probably in oversteer direction
Derived, time variant	Yaw rate with cumulated manipulation:	Velocity:	Side-slip shifted to more oversteer:
	+ Continuous and consequent, mainly feedforward intervention	+ Can be kept as the driver demands	– Unwanted risk of stability loss

From this, the requirement for a consequent, tolerant and mainly control-based operation arises. Unfortunately, a typical TV capable powertrain layout allows only the rear axle to have wheel individual drive torques, so that the TV utilization leads most likely to a more oversteered cornering stiffness balance.

Combining tolerant parameterization and potential oversteer tendency leads to an increased risk of uncontrolled stability loss, which should be avoided, of course. It was previously mentioned that the supervisor wheel torque limitations are inputs of the TV control. Via this interface, an additional function can help to cover this contradictory situation and take over the recognition and control of sudden oversteer situations. This is done assuming a fault tolerance time of 200 to 500 ms for stability loss according to safety concept, as a definition of "sudden". Less aggressive oversteer tendencies can be recognized by observers used by TV and its control can be adapted early enough.

This supervisory function should neglect signals, which are not up to date in the moment of stability loss. Judging the oversteer risk based on side-slip angle velocity, which is directly influenced by TV, enables this direct detection and intervention. Furthermore, such a strategy also allows an easier realization of an ASIL-D capable algorithm, which is most likely a must have requirement according to expected safety goals.

Suppressing side-slip angle velocity must perform well both on surfaces with high and low friction values. In the case of a high friction value, even high lateral dynamics changes can be carried out with stable vehicle behavior, and unnecessary intervention to reduce the sideslip angle velocity can be very disturbing. Parametrizing this function for high friction value cases can guarantee that these situations are not impaired. On surfaces with a medium and low friction value, an already developing skidding tendency can still be recognized and cumulating more than 3° to 5° side-slip angle can be avoided. Getting over this limit may lead to the situation that lateral drive torque distribution as an intervention is not able to stabilize the vehicle anymore and wheel individual braking intervention will be needed.

The suppressing itself can happen via the before-mentioned wheel torque interface to TV. The control, which is based on the side-slip angle velocity, might not cover the use case where the vehicle is slowly building up a side-slip angle over 3° to 5°. A parametrization that allows enough tolerance to not intervene on high-friction surfaces can cause this. In these cases, the TV's observer functions should have sufficient time (between 500 and 1000 ms) to recognize such a situation and modify the TV control consequently.

3 MIL-HIL TESTING OF THE TORQUE VECTORING FUNCTION

To perform dynamic testing of the TV function, the control algorithm was integrated in a Model-in-the-Loop (MiL) and a Hardware-in-the-Loop (HiL) test environment but also on testbed for electric axles. All these test environments were supported by the AVL-VSM simulation environment. In the MiL environment, the TV function was integrated simply into the simulation toolchain. The HiL environment contained the Vehicle Control Unit (VCU) with integrated TV function to further test the behavior of the software on the VCU. On the e-axle testbed, AVL's electric drive unit (e-axle) was built up – including inverter and VCU, on which the TV function was running in an embedded environment. Having real electric motors and the TV algorithm running in a real control unit on the testbed ensures that E/E architecture (for example due to communication delays) and actuator resulted limitations are being considered. Additionally, there is the possibility to validate powertrain simulation in the MiL/HiL environment.

Fig. 3 shows the results of a "Constant speed and increased steering wheel angle" simulation, which was performed on the MiL and HiL test environment.

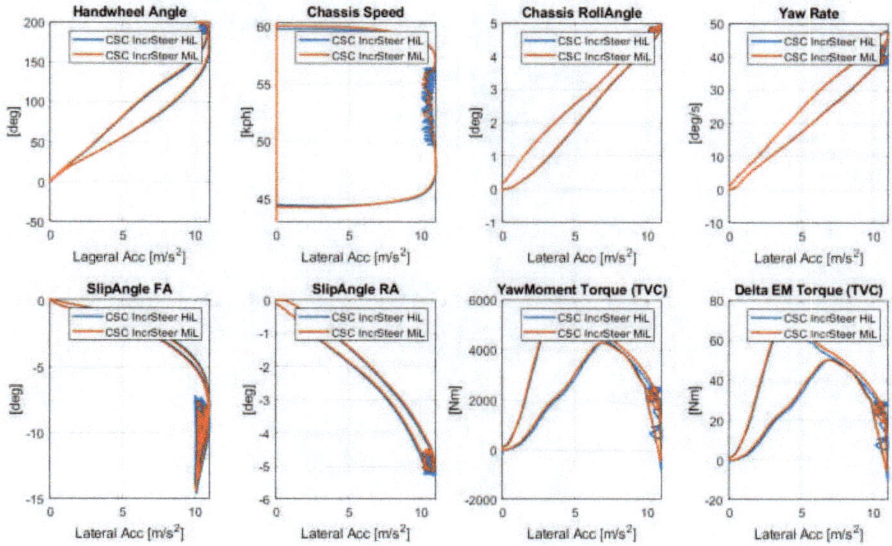

Fig. 3. Simulation results for constant speed and increased steering wheel angle (CSC: Constant Speed Cornering, IncreSteer: Increased Steering, lateral acc.: lateral acceleration, FA: Front Axle, RA: Rear Axle, TVC: Torque Vectoring Control, EM: Electric Motor) (© AVL List GmbH)

Depending on the driving situation, a yaw moment is requested by the TV function. Based on simulation results of the MiL and HiL application, the accuracy between the different toolchains is illustrated. Furthermore, it proves that the development and testing process of a TV function can be performed with a MiL toolchain, which has low setup and operation costs including very fast simulations. Fig. 4 shows the assembly of the testbed for TV function testing. It consists of AVL's e-axle (test specimen) (1), two dynamometer electric motors (2), mounting brackets (3), and measurement equipment (4).

Fig. 4. Testbed assembly (see text for numbers) (© AVL List GmbH)

4 VEHICLE VERIFICATION

AVL's demonstration vehicle is a modified electric passenger car of the type Tesla Model S with AVL's twin-motor electric drive unit on the rear axle. It is pushed in a power-sliding situation on a wet handling course before a chicane on the AVL Zala-Zone test track in Zalaegerszeg (Hungary). Fig. 5 shows the measurement results for this.

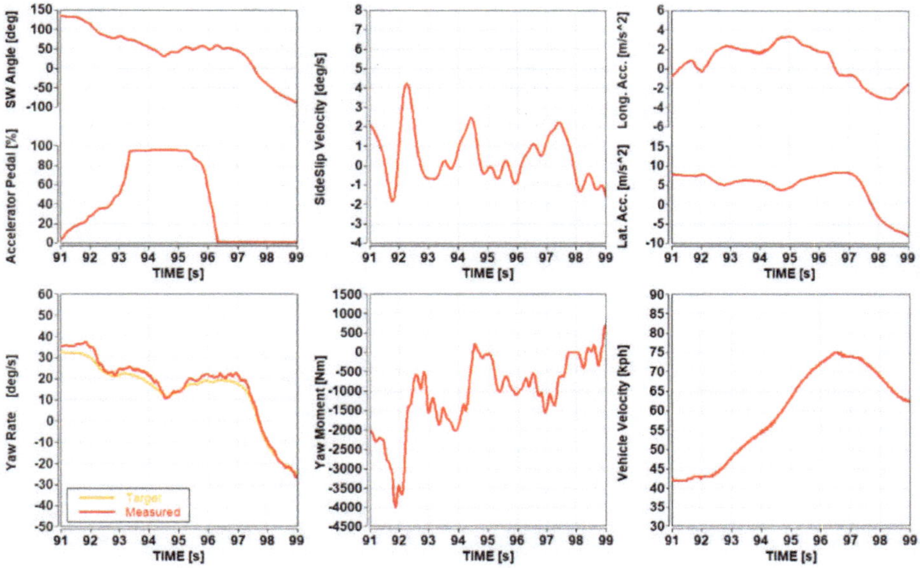

Fig. 5. Vehicle measurement with a Tesla Model S for oversteer compensation on wet handling course (APP: accelerator pedal position, Lateral acc.: lateral acceleration, Longitudinal acc.: longitudinal acceleration) (© AVL List GmbH)

The observer algorithms (typically from the friction value observer) lead to a target yaw rate which is continuously lower than the actual (measured) yaw rate. The TV is acting continuously "against" the driver to stabilize the vehicle. However, as the driver starts to increase the Accelerator Pedal Position (APP), and the side-slip angle starts to turn into the oversteered direction (between 91.5 and 92 s, with the side-slip velocity reaching -1.9 °/s), the yaw moment must be increased even more (up to -4000 Nm) in order to stabilize the vehicle. Thanks to this direct intervention, a controlled power-sliding situation can be realized, and even the second half of the chicane driven through.

References

1. Lenzo, B.; et al.: Yaw rate and sideslip angle control through single input single output direct yaw moment control. In: IEEE Transactions on Control Systems Technology 29 (2020), No. 1, pp. 124-139
2. Hankovszki, Z.; Loof, J.: ATZ-Webinar: Mehr Fahrspaß dank optimierter, integrierte Fahrdynamikregelung. 1. Juni 2023, Online: https://www.springerprofessional.de/atz-webinar-avl-fahrspass/25448036, access: February 20, 2024
3. Hankovszki, Z.; Loof, J.: Targeted driving fun. Lecture, chassis.tech plus, June 5, 2024

Alternative Input Devices in Steer-by-Wire Vehicles

Thomas Maier, BSc

thyssenkrupp Presta AG, 9492 Eschen, Liechtenstein

thomas.maier1@thyssenkrupp-automotive.com

Abstract. The introduction of steer-by-wire technology, which no longer requires a mechanical connection between the steering wheel and steering gear, opens up many new possibilities in the automotive industry. Besides the many possibilities and advantages regarding traffic and accident safety as well as driving experience, this technology also offers the possibility for alternative input devices instead of the classic steering wheel.

This study deals with the possibility of using alternative input devices in vehicles instead of the steering wheel. The focus is on the development and testing of various input devices with force feedback.

Different concepts of micro-steering wheels as well as a concept of a joystick were designed and manufactured, and tests were carried out on a simulator. On this simulator, initial tuning of the steering feel was performed, and the required steering angle ranges correspondingly to the anatomical limits of the various concepts were evaluated. For the dimensioning of the individual components, the occurring steering torques were also investigated. The fully developed concepts were implemented in a steer-by-wire vehicle with an adapted center console, and each has been individually tuned for steering feel on a test track.

In different driving tests with all these various concepts, a subjective as well as an objective comparison between each other and with the classical steering wheel have been performed. Advantages and disadvantages regarding controllability, comfort, and user experience design as well as the limits of the applications and the possible use cases become clear.

Keywords: HMI, steer-by-wire, steering, input devices.

1 Introduction

In recent years, various demands from automotive manufacturers with regard to comfort, the environment and safety have led to major changes in steering system technologies. Hydraulic steering is slowly being displaced from the market by electromechanical steering, and new technologies such as steer-by-wire (SbW) are entering series production at the first manufacturers. Alternative input devices create new possibilities in terms of interior design, primarily with regard to the driver's area. Espe-

P. E. Pfeffer (Ed.), *15th International Munich Chassis Symposium 2024*,
Proceedings, https://doi.org/10.1007/978-3-662-71064-7_8

cially with a view to autonomous driving, the question arises as to what role the steering wheel will play in the future. The higher the degree of automation, the less the driver will have to operate a steering wheel and can possibly devote himself to other tasks. Alternative interior designs, thanks to which the driver's seat can also be used as an office space, for example, can already be seen in many vehicle studies today. [1]

1.1 Motivation

With the market launch of the first SbW systems without mechanical fallback system and semiautonomous vehicles already in production, more advanced autonomous driving and so a realistic demand for alternative interior design and therefore alternative input devices is in sight.

During the continuous further development of various SbW systems in the Department of Research and Advanced Development at thyssenkrupp Steering (TKS), several projects relating to alternative input devices have already been researched and implemented. The first were pure input devices without force feedback, such as a tablet installed in the vehicle, which can be used to control the vehicle from the passenger seat without the steering wheel moving. The latest project implemented was a joystick with force feedback, as part of a master's thesis [2]. This joystick is based on a hardware off the shelf and was first tested on a test bench and later in the vehicle to gain initial experience in this area. The output from this project is promising and the potential of alternative input devices is evident, which is prompting new ideas. As so-called microwheels are already available for people with disabilities, there is great interest in exploring their potential and implementing them with our own hardware and controls.

Previous projects have usually only been superficially embedded in the vehicle environment, which is why their potential has never been fully explored and it is hard to compare them to each other. The complete implementation of different input devices in the same SbW demo vehicle enables better research internally as well as for customer clinics, and also the opportunity to present them to potential customers at demo events.

1.2 Objective

The aim of this research is to develop various concepts for microwheels as alternative input devices, as well as an adapted version of a joystick and to implement, tune and evaluate them, to gain knowledge in the area of alternative input devices and their possible usage.

To enable comparability across the various input devices and with the standard steering wheel, these concepts must be based on existing hardware and software. The devices should be implemented in the existing vehicle environment without compromising on positioning or ergonomics.

An evaluation should show the strengths and weaknesses of the different variants, as well as the possible areas of usage and their limits at the current maturity level. An outlook on the further development or tests with suggested solutions for possibly found weaknesses should be given.

2 Input Devices

In the following, an overview of the most important tuning parameters, an insight into the initial investigation regarding microwheels by using a simulator, and the hardware selection will be provided. Thereafter, the three concepts that have prevailed will be introduced and how they are tuned in terms of the angular operating range.

2.1 Basic Tuning Parameters

In principle, the final devices used for this research will have all the functions that are also used in the current SbW concept. For the better understanding of this study and the preliminary investigation, an overview of the basic parameters is given.

Virtual Gear Ratio (VGR). The VGR describes the ratio between input device angle and road wheel angle in a matrix as a function of steering angle and vehicle speed. It is the tool to define the operating angular range and has the most influence on sensitivity of the steering inputs.

Base Steering Effort (BSE). The BSE function provides the base torque which indicates the shape and the amount of the steering torque but does not directly reflect the torque applied to the device. Torque is added to or subtracted from this base torque via a hysteresis function (HYS), which provides a hysteresis-like steering feel to simulate the hysteresis behavior of different chassis and steering components.

Damping (DAM). The main role of the DAM function is to decrease the kinetic energy of the system when high steering speeds occur. But it can also be used to provide a better damped steering feel for standard driving situations.

2.2 Preliminary Investigation

The aim of the preliminary investigation is to analyze various concepts of microwheels, perform initial tunings, and to estimate the torques and angular operating ranges occurring for the concept designs and hardware selection.

Simulator Session. The simulator used for this investigation is a custom construction by the Research and Advanced Development Department at TKS, which is primarily used for function development for SbW systems.

For the initial investigation, a feedback actuator module (FBA) with a transmission and a maximum torque of more than 10 Nm from a SbW project, was mounted vertically on the base plate, fitted with an adapter for different input devices.

The concept of a microwheel in the form of a rotating plate with a handle, as shown in Fig. 2 on the test setup, is already similarly offered for vehicles adapted for disabilities by various manufacturers without or with a simple force feedback concept [3]. Based on this, various concepts of microwheels with different diameters, handles, knobs, and other control variants were investigated on this setup. Initial tunings for these concepts were tuned on the simulator, with the microwheels as input devices and a virtual driving simulator software.

Torque measurements were performed during various maneuvers to determine the maximum torque and the angular operating ranges correspondingly to the anatomical limits were defined. For optimal ergonomics also the positions of the various knobs and handles were investigated.

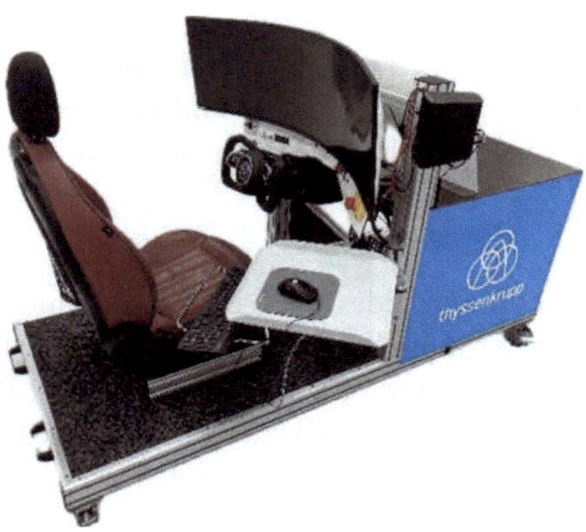

Fig. 1. TKS SbW Simulator

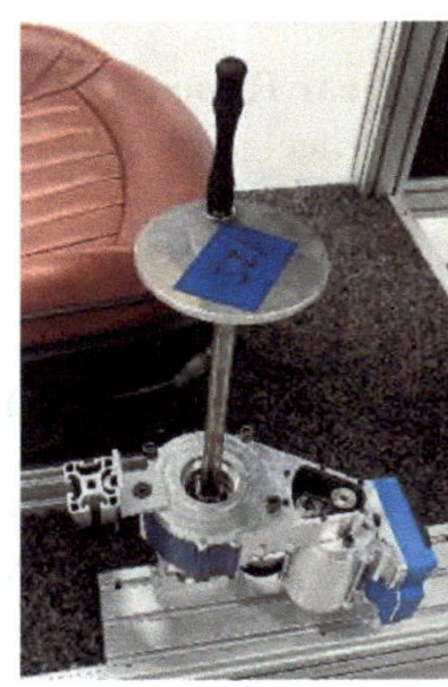

Fig. 2. Microwheel-Test-Setup

Hardware Selection. The investigations on the simulator showed a maximum torque of 2.5 Nm in maneuvers hitting the end-lock position and a torque range of up to 1 Nm during daily driving maneuvers. To cover these torques and have enough headroom, the decision was made to use a motor from an existing FBA module as a direct drive without the need for a transmission as for a conventional steering wheel is used. The choice fell on a three-phase synchronous motor with 210 W power and a maximum torque of 6 Nm. In addition to its performance data, it can also be well integrated into the vehicle environment, due to its compact dimensions.

2.3 Final Concepts

With the findings of the preliminary investigation and with a view to the fullest possible integration into the vehicle environment, two concepts in the form of a microwheel were developed. To distinguish between the two concepts for this research, the first concept was given the name *Microwheel,* and a concept adapted from it was named *Fingerwheel*, as this is the most compact overall concept and is controlled purely by finger force. In addition to the two concepts, an adapted version of a *Joystick* from a previous study was designed. Beside the concept description, the virtual gear ratio parameters will be discussed for each concept, as they have a big influence on this study.

Microwheel. The basic microwheel concept with a rotary plate and a knob was already presented within the preliminary investigation. As already mentioned, various combinations of plates and handles were tested in the simulator session. Since the version with a knob is a more compact variant and the experiments on the simulator showed no disadvantages, the knob variant was chosen for the final integration in the vehicle. This knob is rotatably mounted on a ball bearing and can be hold with just three fingers.

Fig. 3. Microwheel

Due to the angular operating range of +/- 85°, the virtual ratio could be realized in an almost linear curve across all speed ranges (Fig. 4). The devices are tuned till a maximum speed of 120 kph for this investigation. The solid lines describe the standard operating range in a stable vehicle condition while the dotted lines represent the area which is only reached in non-linear situations. A nearly constant ratio across the steering angle could be achieved for the *Microwheel* in the slow speed area. With higher speed, the ratio in the standard operating range became more indirect to reduce sensitivity. For a constant end-lock position for this concept at all speeds, the ratio is slightly higher in the range of large angles, outside the standard operating range.

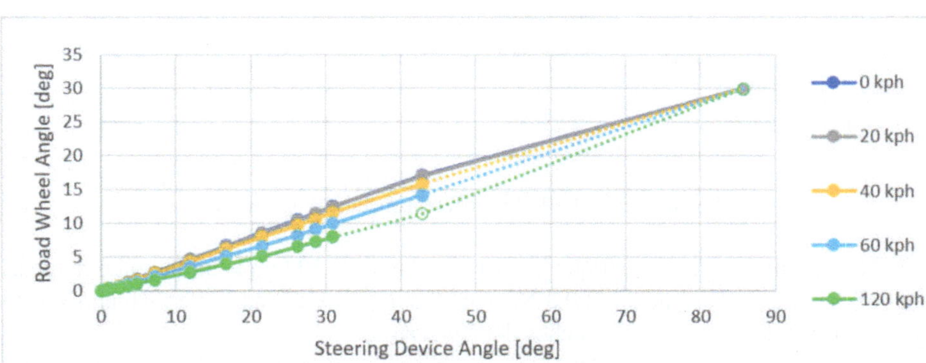

Fig. 4. Virtual Ratio for different speeds, Microwheel

The maximum occurring torque in an 18 m slalom at 50 kph is about 0.3 Nm. The shape of the steering torque is for all devices similar to standard tunings with normal steering wheels but downscaled to match the smaller diameters of the devices and adjusted to the changing virtual ratio.

Fingerwheel. The inspiration for this concept was, on the one hand, the classic remote control as used for remote-controlled model cars and, on the other hand, the fact that similar control elements are already used for infotainment, such as BMW's iDrive [4]. This concept is therefore the most compact and can be integrated most smoothly into the vehicle environment. To improve the ergonomics of the device, in addition to a horizontal aligned device as known from infotainment control elements, a version tilted to the side was also designed, as this is closer to the natural orientation of the human hand.

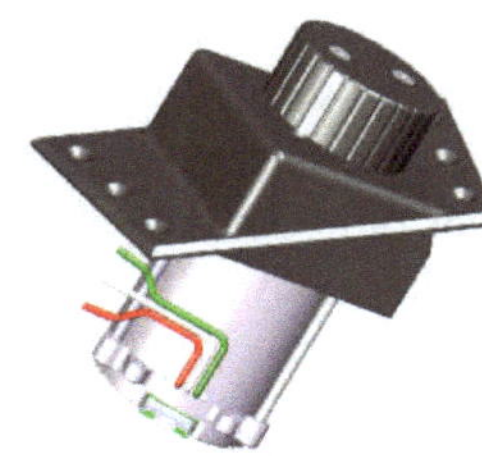

Fig. 5. Fingerwheel, tilted version

Preliminary tests have shown an operating range of +/- 40°, with operating within 30° being the most comfortable. To fully utilize the potential of the variable transmission ratio in this small range, a very high and non-linear virtual ratio is used for this device in the lower speed range (Fig. 6). To reduce sensitivity at higher speeds and provide an almost linear behavior in the standard operating range, the end-lock position has been moved out of the ideal 30° operating range. The non-linear area above 30° will not be reached under linear driving conditions but can still be reached in an emergency maneuver without repositioning the hand.

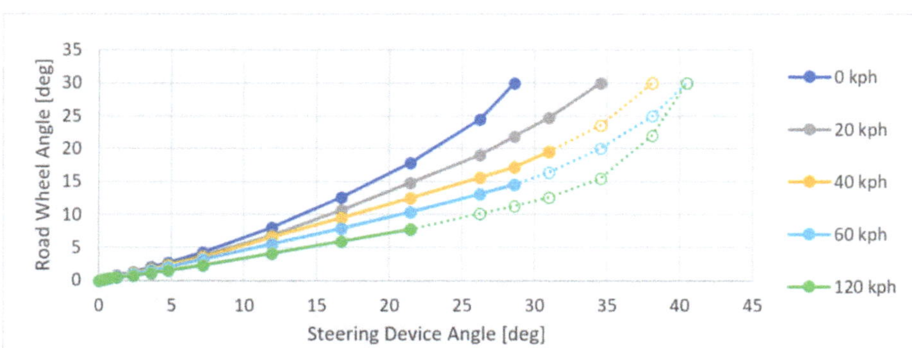

Fig. 6. Virtual Ratio for different speeds, Fingerwheel

The maximum occurring torque in an 18 m slalom at 50 kph is in the same range as that of the *Microwheel*, at about 0.3 Nm, but the hysteresis is narrower, which leads to a springier feeling and better matches the *Fingerwheel*.

Joystick. The concept of the *Joystick* is a further development from tests carried out as part of a master's thesis at TKS [2]. The integration of the *Joystick* turned out to be quite difficult, which is why this concept uses a belt to separate the feedback motor from the control element to position it as ergonomically and space saving, as possible in the vehicle. The *Joystick* is aligned vertically in such a way that it is controlled by a rotational movement of the forearm via the wrist (Fig. 7). For the vehicle tests the concept was adapted as a direct drive, as in this vehicle there was enough space, and the realization would have exceeded the time frame. However, this did not influence the investigation.

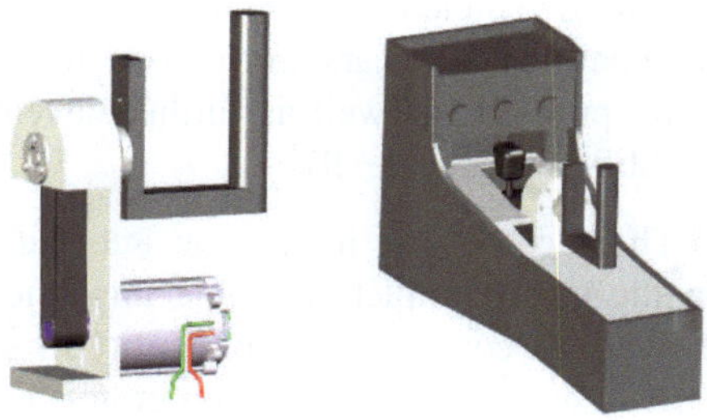

Fig. 7. *Joystick* (left), *Joystick* integrated (right)

The possible angular operating range for this concept was set at +/- 55° due to the experience gained with the previous joystick-concept. During the tuning session it became clear, that this rotational movement of up to 55° quickly leads to fatigue in the forearm. For this reason, the possibility of variable transmission was fully utilized for the tests and the operating range, especially in the low-speed range, was kept as minimal as possible. At higher speeds, the full range is used, as the high angles are only reached in emergency maneuvers (Fig. 8).

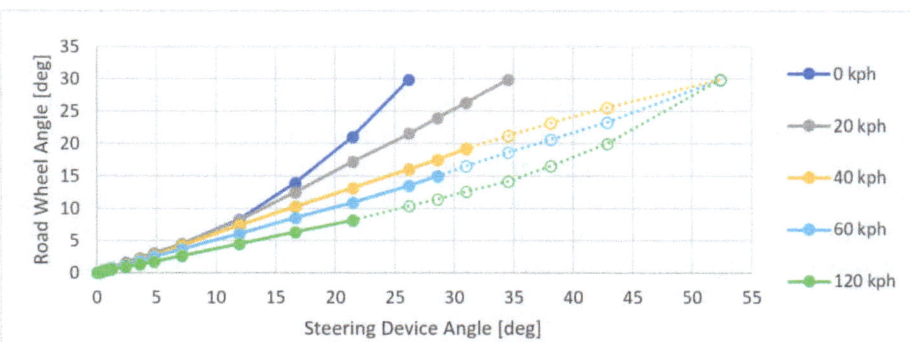

Fig. 8. Virtual Ratio for different speeds, *Joystick*

The maximum occurring torque in an 18 m slalom at 50 kph is about 0.55Nm and therefore higher than for the other devices, but the *Joystick* is controlled in a way that allows higher torques.

2.4 System Integration

The used vehicle for this investigation is one of the company research cars and is used for testing and demo purposes as well as further development of the SbW system (Fig. 9).

Fig. 9. SbW Golf 7

The focus of the integration into the vehicle is on the most complete possible implementation of the various concepts and the ability to swap them quickly and easily without compromising on ergonomics. To ensure this, the transmission control module was moved forward by 120 mm and the shift lever was shortened. The three devices share an ECU, which is permanently installed in the center console under the control panel to activate the system and is connected to the devices via a plug. They are all mounted with 6 bolts into an adapter plate in the center console as seen at the example of the *Fingerwheel* in Fig. 10, what makes it very easy to swap them.

Fig. 10. Center console with Fingerwheel

3 Test Maneuvers

The aim of the selected test maneuvers is to evaluate the steering behavior of the various devices in terms of controllability, fatigue, sensitivity, and user experience design. Therefore, the maneuvers are split into driving situations from low dynamic to high dynamic to better analyze the different areas. All the maneuvers were carried out and evaluated subjectively by a professional test driver from TKS. As a base reference for the evaluation, all maneuvers were carried out with the normal steering wheel in addition to the three devices. From this base reference, the particular performances were rated from significantly lower to significantly higher, each with three levels of subdivision. Selected signals were measured, logged, and analyzed for all the maneuvers.

3.1 Low Dynamic / Daily Maneuvers

Line Following. Following a predefined track at a set speed of 20 kph with the aim of not leaving the marked line with the left front tire over five laps. One warm-up lap.

Reverse Parkour. Driving backwards through a course set with cones with the aim of not touching any cones over three laps. One warm-up lap.

Parking. Reverse parking and side parking into a parking space defined by cones. Three executions per maneuver.

3.2 Medium Dynamic / Basic Maneuvers

Handling Course. Reproducible drive around the well-known handling course on the racing line with moderate speed and without pushing the limits. One warm-up lap.

Figure-Eight. Circle-radius of 8.5 m for each circle; const. lateral acceleration while cornering of 6 m/s^2; 5 laps; With the aim of a smooth transition between the two circles and not touching any cones.

Sweep Steer. Sine on straight road; target acceleration: +/-1 m/s^2; const. vehicle speed: 60 kph; frequency: 2 Hz; 12 cycles.

Fatigue Circle. Constant circling at 60 kph and 100 m radius over four laps with a sweep steer overlapping in laps 2 and 4.

3.3 High Dynamic / Emergency Maneuvers

Double Lane-Change. ISO 3888-1; steady throttle position; const. vehicle speed: 50 kph, 60 kph, 70 kph; Three executions for each speed.

Skid Plate. Random skidding direction; steady throttle position; entering vehicle speed: 50 kph; no obstacle avoidance; Six executions.

Slalom. Cone distance: 8 m; lane width: 3 m; cones: 7; const. vehicle speed: 50 kph;

Mu-Split Braking. Low friction surface for left tires, high friction surface for right tires; entering vehicle speed: 60 kph;

4 Results Vehicle Tests

The results of these maneuvers were divided into controllability, fatigue, sensitivity, and user experience design. An overview of the ratings and a detailed analysis for each test can be found in the following chapter. In the graphic illustrations, the devices are referred to by the following initials: MW (Microwheel), FW (Fingerwheel) and JS (Joystick)

4.1 Controllability

For a better overview the results regarding controllability are divided into subsections from low to high dynamic.

Low Dynamic. The controllability of the low dynamic maneuvers is generally quite good, as visualized in Fig. 11. All three devices felt very sensitive at low speeds, which is described in more detail in a later sub-chapter, but this had no influence on controllability and did not affect precise line following. Weaknesses were only evident here with the *Microwheel*, as the controllability is very much influenced by the angle-dependent position of the knob due to ergonomic reasons. With the *Fingerwheel* as well as with the *Joystick*, there were no subjective difficulties in following the line over all necessary steering angles compared to the original steering wheel. The evaluation of the trajectory data and the steering angle signal shows no significant deviations from the steering wheel in terms of line deviation and significant steering angle corrections for all three devices during the Line Following Maneuver.

When driving backwards through the set-up course, the controllability with all three devices was just slightly lower to the previous maneuver, but for ergonomic reasons you are forced to drive purely using the mirrors (or via the rear camera), which leads to a certain restriction, depending on the driver's preferences.

The effect of the variable ratio comes into action when parking. As there is no need to reach around the steering wheel and angle correction is realized more effective, parking in reverse and sideways is slightly easier than with the steering wheel. Subjectively, it is easier to focus on the positioning of the car.

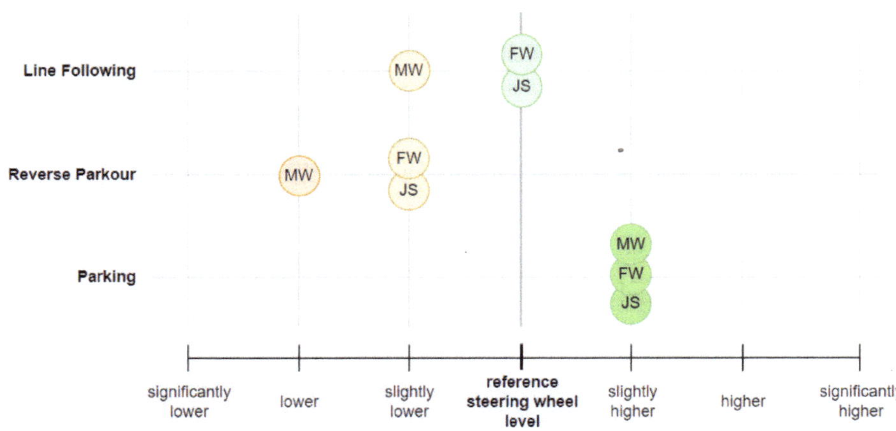

Fig. 11. Controllability low dynamic maneuvers, subjective ratings

Medium Dynamic. Through the Handling Course as well as the Figure-Eight Maneuver, the vehicle could be controlled very good with the *Fingerwheel,* and the *Joystick* and it could be positioned very well on the ideal line on the handling track. With the *Microwheel,* similar to the Line Following Maneuver, it was very dependent on the current position of the knob how easy it was to hold the line. As a result, the turn-in points and the desired apex were missed several times and a lot of corrective steering was necessary during cornering on the handling track. During the Figure-Eight Maneuver it was slightly more difficult to find the correct angle for the turn in after the redirecting point with the *Microwheel* compared to the other devices.

The Sweep Steer maneuver could be performed very precisely with the *Microwheel,* whilst these very small, precise steering deflections are slightly more difficult with the *Fingerwheel.* The Fatigue Circle Maneuvers where the sine deflection is added during cornering a constant circle, could be performed precisely with the *Fin-*

gerwheel, while it was slightly more difficult to keep the desired lateral accelerations reproducible with the *Microwheel*. There were also differences with the *Microwheel* depending on the direction of the Fatigue Maneuver as it was slightly more difficult to perform the maneuver precisely in an anti-clockwise direction, which can again be explained to the different positions of the knob while cornering to the right or left side. In principle, the controllability over all maneuvers with the *Joystick* is very good, but it decreases over time due to fatigue.

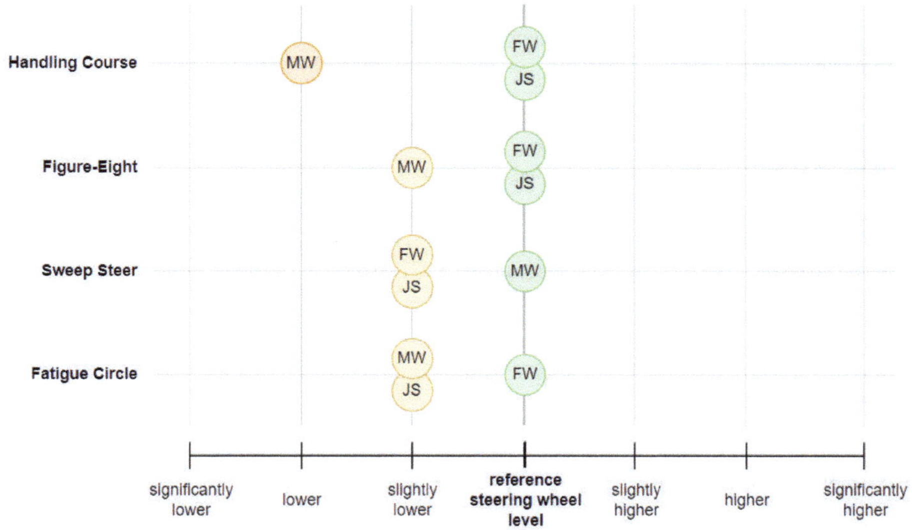

Fig. 12. Controllability medium dynamic maneuvers, subjective ratings

High Dynamic. The Double Lane-Change Maneuver became more difficult as the speed increased. At 50 kph, the vehicle could be maneuvered through the cones very effortlessly and in a controlled manner with all devices. With the *Microwheel*, it was no longer possible to bring the vehicle back to the initial lane at 60 kph without hitting several cones and at 70 kph, the vehicle was no longer controllable after the first steering maneuver into the offset lane. With the *Fingerwheel* and the *Joystick* there were no limitations up to 60 kph and the vehicle could be steered very precisely and effortlessly through the course. At 70 kph the controllability with the *Fingerwheel* was lower and the maneuver could not be executed without hitting cones in the last sector of the double lane-change course. With the *Joystick* the highest possible speed of the three devices without hitting a cone was 65 kph but at 70 kph the controllability was less than with the steering wheel and it was not possible to avoid all cones.

During the Skid Plate Maneuver, the vehicle could be controlled and maneuvered as good or better as with the steering wheel with all three devices. With the *Fingerwheel* and the *Joystick* the vehicle was brought back under control even more controlled than with the steering wheel as the first steering input is very effective. The precision of the first input felt even higher with the *Joystick*, so the vehicle was slightly easier to control than with the *Fingerwheel*. With the *Microwheel*, the effect of fast counter-steering also came into play, but compared to the other devices, it was more difficult to hit the right angle precisely in order to fully utilize the advantage.

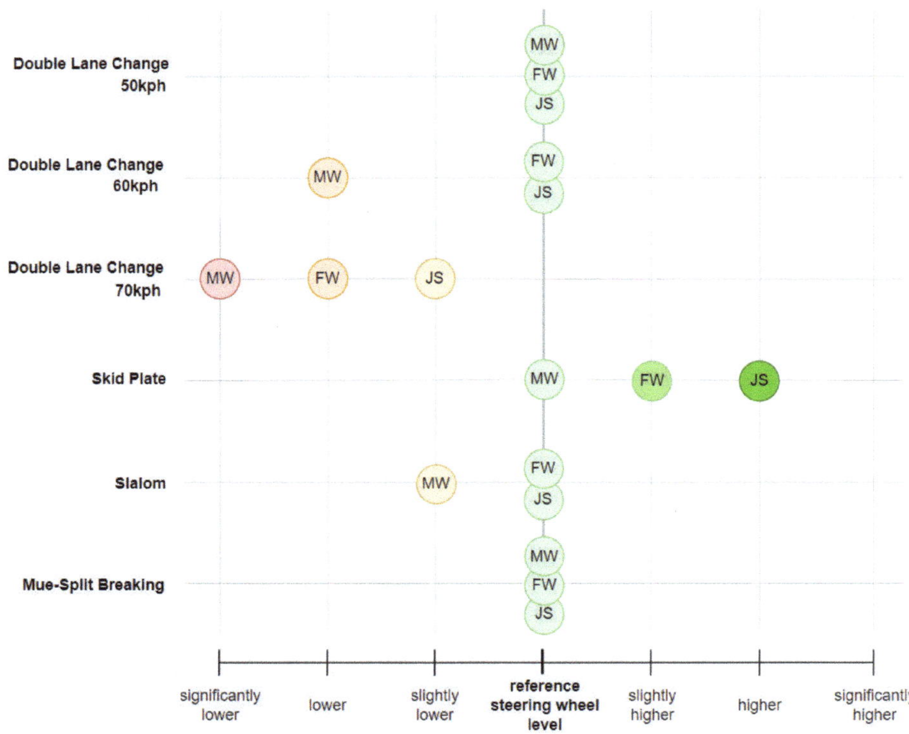

Fig. 13. Controllability high dynamic maneuvers, subjective ratings

The analysis of the measurement data confirms the subjective feeling of the effectivity of the first steering input (Fig. 14). The acceleration and the average speed of the steering rack and road wheel angle while counter-steering were higher than with the steering wheel, which led to smaller required road wheel angles and better control during oversteering. The following graphic shows an example of one maneuver per device and direction of deflection. In addition to the described characteristics, it can also be seen that these effects are better when counter-steering to the left, which could be due to the ergonomic differences between left and right steering with these devices.

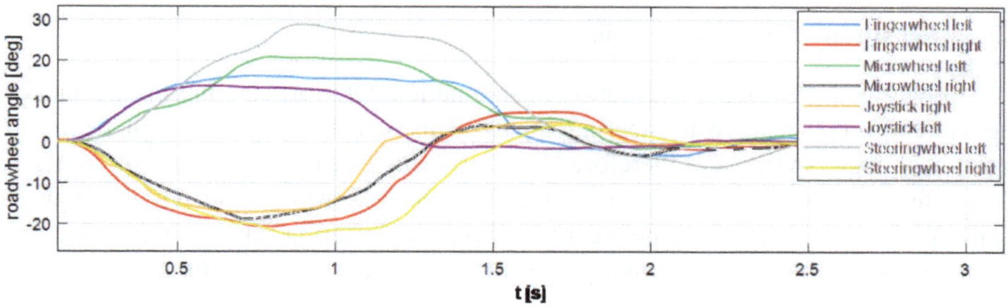

Fig. 14. Comparison Skid Plate Maneuver

The slalom as well as the Mu-Split Breaking Maneuver could be executed well and precisely with all devices, although the precision of the execution of the slalom with the *Microwheel* was a little more difficult.

4.2 Fatigue

In terms of fatigue during the low-speed maneuvers, the *Joystick* showed a disadvantage, whereas the *Fingerwheel* has advantages here, not least because of the lower torque. In the medium dynamic tests with higher steering torques, the gap between the devices increased. In the Fatigue Circle Maneuver, which aims to find the limits of the devices, the *Joystick* performed significantly worse and higher fatigue in the forearm was also observed during handling and the Sweep Steer Maneuver. The *Microwheel* and the *Fingerwheel*, on the other hand, also showed advantage over the steering wheel, which were most noticeable in the Fatigue Circle Maneuver.

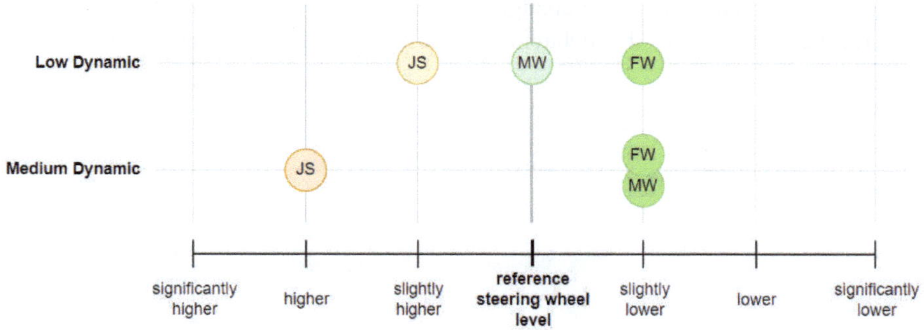

Fig. 15. Fatigue, subjective ratings

Since the high dynamic maneuvers do not last long, the aspect of fatigue is not evident and therefore was not evaluated.

4.3 Sensitivity / Vehicle Reaction

At low speeds the higher gear ratio in combination with low steering torque, makes the system very sensitive to steering inputs. This sensitivity led to high rack and roadwheel angle speeds and therefore to a high vehicle reaction and loss of comfort. This effect was most noticeable with the *Fingerwheel*, which also has the highest VGR. With more dynamic maneuvers and increasing steering torques, this effect became less noticeable, and with highly dynamic maneuvers it was hardly present.

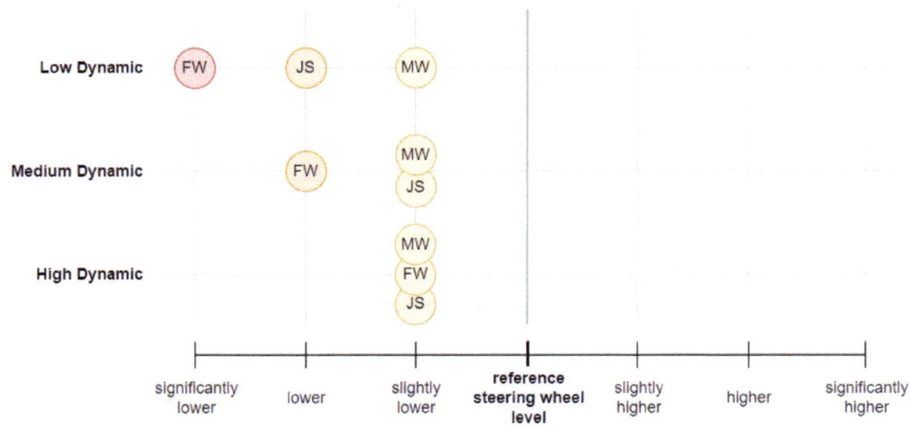

Fig. 16. Comfort due to sensitivity, subjective ratings

An analysis of the measurement data showed that the roadwheel angle speeds with the *Fingerwheel* were sometimes twice as high as with the steering wheel during the Line Following Maneuver when changing direction in corner combinations (Fig. 17). Also, for the other devices this behavior is clearly noticeable in the measurement data.

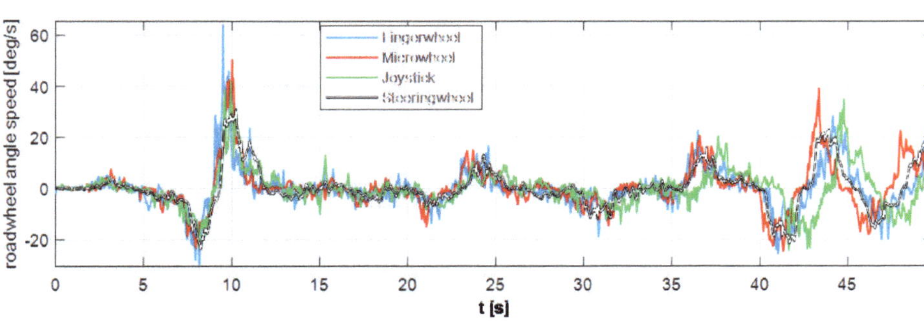

Fig. 17. Comparison of the road wheel angle speed at line following

Besides the high roadwheel angle speeds during the steering maneuvers themselves, a frequency spectrum analysis of the signal also shows high steering angle velocities in a frequency range of 1-10 Hz. On the one hand this can be assigned to very small angle corrections, but also shows that even the slightest insecurity or shaking of the hand is passed on directly to the wheels due to low base torque and damping of the steering devices combined with the high virtual ratio.

4.4 User Experience Design

Regarding usability and implementation in the vehicle environment, the devices are evaluated in three aspects from low to high.

It took very short time to get used to the *Fingerwheel* and the *Joystick*, steering with these devices felt very intuitive. Due to the position-dependent different movement when turning the *Microwheel*, it took a little longer to get used to it.

In terms of ergonomics, the *Fingerwheel* and *Joystick* have the advantage that the arm can be placed on the armrest at almost any time, as the movements to control the devices come from the forearm/wrist and fingers. Controlling the *Microwheel* is not less comfortable as turning a steering wheel, but it involves more than just the forearm and wrist, which means that the arm cannot be placed as comfortably as with the other devices.

There are major differences in the integration of the devices. As the *Fingerwheel* with its compact size can be integrated very well into the center console, it would not require much adaptation. The *Microwheel* and especially the *Joystick*, on the other hand, require more space and cannot be integrated as well as the *Fingerwheel* in terms of space and design.

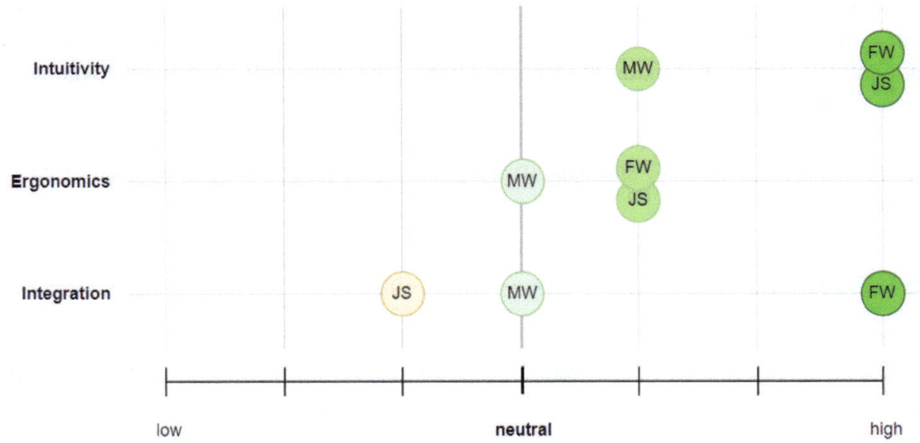

Fig. 18. User Experience Design, subjective ratings

5 Summary

An overview of the main findings per device as well as potential solutions for improvement of the main issues can be seen in the following summary.

5.1 Microwheel

The *Microwheel*, which is the basic idea concept, performed lowest overall in the comparison. In terms of controllability, it showed disadvantages compared to the other devices in all tests, especially during the Handling Course and the Double Lane-Change Maneuver. The angle-dependent knob position had the biggest influence on the lower results, as it does not result in a homogeneous and slightly unpredictable movement flow. Still, also the *Microwheel*, as well as the other devices showed advantages over the conventional steering wheel when it comes to parking maneuvers. Regarding fatigue, the *Microwheel* performed quite well and in terms of vehicle reaction it was able to perform best, primarily due to its low virtual ratio also at lower speeds. While it is limited in terms of integration due to its larger size, it is also behind the other two devices in terms of ergonomics and intuitiveness.

5.2 Fingerwheel

Regarding controllability, the *Fingerwheel* showed that there were no significant compromises in the low and medium dynamic area, only reversing was slightly more difficult compared to the conventional steering wheel. In the highly dynamic maneuvers, there were limits for the Double Lane-Change Maneuver at high speeds, but advantages on the skid plate. In terms of fatigue, the *Fingerwheel* showed advantages overall, especially compared to the *Joystick*. As it is the most sensitive of all devices, the vehicle reaction to steering inputs is too high, especially at low speeds. The *Fingerwheel* is very intuitive and user-friendly and integrates best into the existing environment.

5.3 Joystick

The *Joystick* achieved the best results in terms of controllability, especially with highly dynamic maneuvers. It has limitations when reversing in common with the other devices. Fatigue over time was found in all areas, most notably in the medium dynamic area. The high vehicle reaction with steering inputs, especially at low speeds, was also evident with the *Joystick* due to the very high virtual ratio, whereby the higher torque helps to counteract this slightly. In terms of integration into the interior environment, the *Joystick* is the most difficult to integrate due to its size and orientation.

5.4 Main Issues and Potential Solutions

The investigations regarding sensitivity showed that the defined operating range in the preliminary tests for the *Fingerwheel* and *Joystick* under normal driving conditions were not fully utilized while the tests and the actually used range during standard operations was a lot smaller. On the other hand, an increasing ratio towards the end stop was not considered disturbing in certain situations.

This opens up the possibility of increasing the operating ranges and utilizing the entire range, especially at low speeds, which enables a much more indirect virtual ratio, particularly in the low-speed area.

The results regarding fatigue indicate that higher steering torques would be possible for both the *Fingerwheel* and the *Microwheel*, which would also counteract the sensitivity at low speeds and possibly also enable more control in the area of the Double Lane-Change Maneuver. On the other hand, the base torque of the *Joystick* could be lowered to reduce fatigue.

Another effective method of reducing sensitivity in all speed ranges could be to significantly increase the damping at high steering input speeds. This could reduce the high roadwheel angle speeds and also result in more controllable behavior during avoidance maneuvers.

These modifications could also reduce the high-frequency movements of the rack or road wheel, providing additional comfort and protecting the system from stress. In addition, a function that filters the input signal could also help and should be investigated in further steps.

6 Conclusion

Three different concepts were successfully developed and partly seamlessly implemented in the interieur with the option of switching between them relatively quickly and easily.

Various test maneuvers revealed the advantages and disadvantages of the individual devices. Also, the use of a more direct virtual ratio in the various driving situations, and where there is still potential for improvement has been shown. The current state of development still shows limits in terms of comfort at low speeds and controllability in highly dynamic avoidance maneuvers which require further investigations.

By analyzing the measurement data, conclusions could be drawn about the subjective evaluations of the various maneuvers, and possible solutions could be identified. In order to test or implement these suggested solutions, a more detailed tuning session focusing on the difficulties found, would have to be carried out. Besides this tuning session, further investigations about the performance of left-handed people, as well as right-handed people in left-hand traffic should be part of a study with a higher number of participants.

Overall, this investigation shows that the potential of alternative input devices is significant, and many possible areas of usage, for example as a replacement for the steering wheel in the field of highly autonomous driving, are feasible.

References

1. ACE Auto Club Europa e.V. Homepage, https://www.ace.de/autoclub/ace-lenkrad/test-und-technik/test-und-technik-artikel/artikel/automobile-zukunft-teil-ii-vom-innenraum-zum-smartroom/, last accessed 2024/02/23.
2. Maximilian Renner, JOYSTICK AS ALTERNATIVE STEERING INPUT DEVICE IN STEER-BY-WIRE VEHICLES, TU Graz / thyssenkrupp Presta AG, Graz (2021).
3. Bozzio Hompage, https://www.bozzio.ch/joysteer/mmis, last accessed 2024/03/20.
4. BMW Homepage, https://www.bmw.com/en/events/idrive/index.html , last accessed 2024/03/20.

Presentation of a Steer-by-Wire Force Feedback Actuator with Direct Drive E-Motor and MR-Brake

Matthias Niegl[1], Johannes Hendewerk[1], Matthias Becker[2], Stefan Battlogg[3], Matthias Holzleitner[4]

[1] MdynamiX AG, Heßstr. 89, 80797 Munich, Germany
[2] Munich University of Applied Sciences, Dachauerstraße 98b, 80335 Munich, Germany
[3] INVENTUS Development GmbH, Montafonerstraße 68, 6771 St.Anton i.M., Austria
[2] STIWA Group, Salzburger Str. 52, 4800 Attnang-Puchheim, Austria
matthias.niegl@mdynamix.de

Abstract. This paper presents a novel design of a steer-by-wire (SbW) force feedback actuator for automotive applications. The actuator consists of a direct drive electric motor (e-motor) and a magnetorheological brake (MR-brake) that are integrated in a compact structure. The e-motor provides the steering feel related torque and a fast response, while the small and energy efficient MR-brake provides passive damping and high end-stop torque. By the combination, the e-motor could be designed smaller, uses less energy and materials. The performance of the actuator combination is evaluated by simulation and driver-in-the-loop tests. The controller design includes a torque splitting algorithm, which separates the feedback torque into MR-torque and e-motor torque. This algorithm has to provide high fidelity force feedback and robust stability under various vehicle dynamic maneuvers. The proposed actuator is suitable for SbW systems that require high performance, safety, and reliability.

Keywords: Steer-by-Wire, Force Feedback Actuator, Magnetorheological

P. E. Pfeffer (Ed.), *15th International Munich Chassis Symposium 2024*,
Proceedings, https://doi.org/10.1007/978-3-662-71064-7_9

1 Current Force Feedback Actuator Designs

Current force feedback actuators which will be released in series production vehicles soon are using the same basic technology as column assisted electric steering systems. A small e-motor generates feedback torque with a worm gear towards the driver. A torque sensor is still required for a consistent steering feel to compensate friction tolerances from the manufacturing or wear [1]. A worm gear is a cost effective, simple and well-known component, the risk of a failing steer by wire system motivates the suppliers to use proven technology. Not all suppliers are focusing on this approach, e.g. Thyssenkrupp introduces a direct drive force feedback technology without a torque sensor [2]. Direct drive e-motors promise a rigid steering feel with the possibility to generate frequencies up to 50 Hz. The drawback which prevents a more common use of direct drive e-motors is the requirement for the end stop torque which is typically higher than 20 Nm. The exact number is still to be discussed among the OEMs. This would result in heavy and large e-motors which are cost intensive and have major drawbacks in terms of packaging. Figure 1 shows close to series production SbW columns from Nexteer Automotive and Thyssenkrupp.

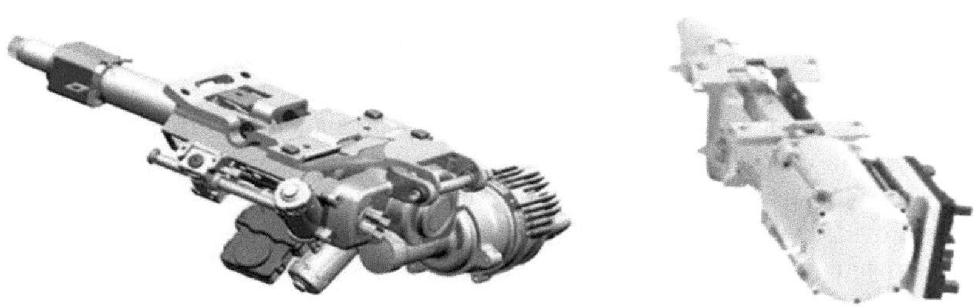

Fig. 1. SbW Force Feedback Actuator/Steering Column from Nexteer Automotive [3] and Thyssenkrupp [2]

There are several possible solutions to using a small direct drive e-motor for functions related to steering feel but still providing high torque output:

- Use the maxed out peak torque of the e-motor
  - \+ No additional parts
  - \+ Modelled end stops
  - \+ Flexible end stop position
  - \+ Fixed steering wheel when entering or exiting the vehicle
  - \- Limited duration of maximum power
  - \- High load on the vehicle's power supply and ECU/power electronics
  - \- High temperatures
  - \- High e-motor torque vs. driver (functional safety aspect)

- Mechanical end stops in series to an e-motor
    - \+ Cost effective
    - \+ High maximum torque
    - \+ Mechanical feel
    - \+ Unlimited duration of maximum torque
    - \+ No consumption of electrical energy
    - \- End stop position and stiffness is not flexible
    - \- No fixed steering wheel when entering or exiting the vehicle

- Mechanical brake in series to an e-motor
    - \+ Cost effective
    - \+ High maximum torque
    - \+ Mechanical feel
    - \+ Unlimited duration of maximum torque
    - \+ Low energy consumption
    - \+ Flexible end stop position
    - \+ Fixed steering wheel when entering or exiting the car
    - \- End stops modeling possibilities only limited
    - \- Slow response and release time
    - \- Additional hardware part
    - \- Wear of the brake components
    - \- Noise during mechanical contact

In summary, this short comparison between different approaches shows there is no simple solution for the direct drive force feedback actuator in existence. A combination of an e-motor with an MR-brake described in the following could solve the tradeoff between good steering feel and reduced costs or packaging.

2 Design and Theoretical Benefits of the Actuator Combination

A magnetorheological brake is a device that uses the properties of a special fluid or powder to control the braking force. The powder, called carbonyl iron powder or MRP, changes its viscosity when exposed to a magnetic field. The brake has two main parts: a rotor and a stator, which are separated by a small gap filled with MRP. The rotor is connected to the shaft that needs to be slowed down, and the stator is fixed to the housing. An electric coil is attached to either the rotor or the stator, and can generate a magnetic field when a current is applied to it. The magnetic field closes from the rotor to the stator through the gap, and it causes the MRP particles to align and form magnetic bridges between the two surfaces. Figure 2 shows a simulation without and with a magnetic field.

Fig. 2. Simulation of the formation of particle chains in an external magnetic field. left: Without magnetic field, right: With magnetic field. [4] (© Fraunhofer IWM)

These bridges create shear stresses that resist the relative motion of the rotor and the stator, thus producing brake torque. The brake torque is approximately proportional to the current in the coil, the area of the gap, and the distance between the rotor and the stator. By adjusting the current, the brake torque can be controlled precisely and smoothly. The magnetorheological brake is a simple, reliable, and efficient way of braking. To reach a counter torque of 25 Nm the MR-brake prototype consumes 60 W electrical power at 12 V [5]. An electric motor wheelbase such the Simucube 2 Pro generates a maximum torque of 25 Nm at 450 Watts peak [6].

Figure 3 shows on the left side the INVENTUS SbW actuator concept dimensions are significantly smaller than a compared gaming wheelbase. This hybrid force feedback actuator includes a 5 Nm e-motor and a 25 Nm MR-brake.

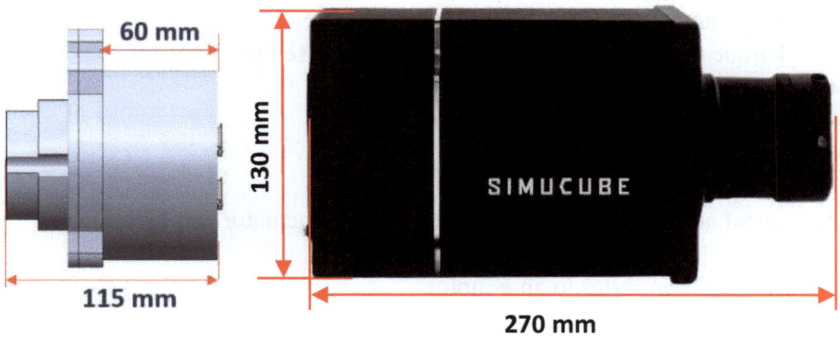

Fig. 3. INVENTUS Steer-by-Wire Force Feedback Actuator Concept Dimensions [5] vs. Simucube 2 Pro [6] (to scale)

The major reasearch question is wether a hybrid force feedback actuator can generate the same quality of steering feel as a pure e-motor. In order to answer that, the hybrid actuator designed for this study (Figure 4) did not focus on size or packaging. The e-motor itself can generate up to 15.7 Nm of continous torque and 30.0 Nm of peak torque. This enables a direct comparison on one system between the hybrid design and the pure e-motor with a disabled MR-brake. If both actuators are enabled, the e-motor torque can be limited to any low peak torque from a close to series production concept.

The actuator further has a MR-brake designed by INVENTUS Development GmbH with a peak torque of 8 Nm. Figure 4 shows the hybrid actuator cross-section and the real actuator including a high precision angle sensor.

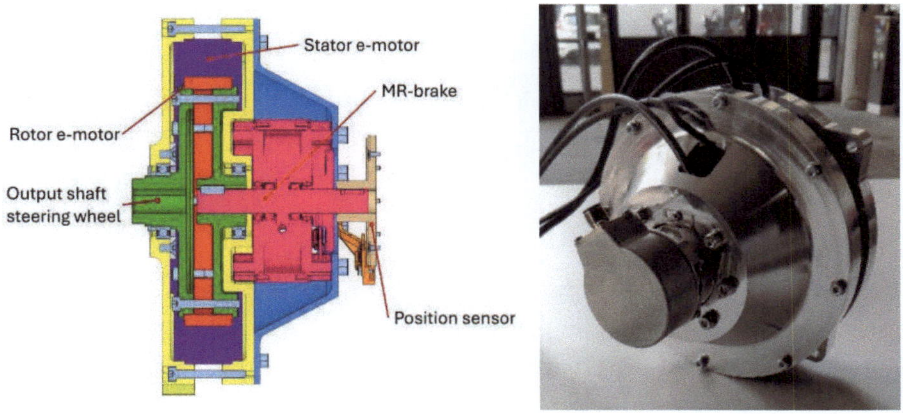

Fig. 4. Steer-by-Wire Force Feedback Actuator with Direct Drive E-Motor and MR-Brake

The overall requirements for a close to series production actuator are more detailed regarding the steering feel e.g.

- Maximum passive friction torque
- Maximum passive damping torque
- Torque up/down ramping torque and response time
- Low torque non-uniformity and ripple
- Stable in automotive temperature ranges
- High cutoff frequency (>30 Hz)

The theoretical advantages and disadvantages of the actuator combination are:

- MR-brake in series to an e-motor
  - \+ Cost effective
  - \+ High maximum torque
  - \+ Mechanical feel
  - \+ Unlimited duration of maximum torque
  - \+ Low energy consumption
  - \+ Flexible end stop position
  - \+ Fixed steering wheel at entering or exiting the vehicle
  - \+ Modelled end stops
  - \+ Low response and release time [7]
  - \- Additional hardware part

3 Controller and Torque Splitter

An MR-brake can only be controlled semi-actively and can therefore only take over two of four quadrants in the torque-velocity diagram during steering. To put it simply, the MR-brake can only brake, i.e. not actively move the steering wheel. An electric motor is required for e.g. active return. Figure 5 shows the diagram of an ideal torque splitter between the MR-brake and the e-motor applied on measurement data. The data used for the analysis represents authentic driving data from a variety of driving scenarios. A clear separation of the ideal areas of usage is evident, and wherever feasible, the e-motor has been substituted. With this strategy, the MR-brake is used effectively as often as possible.

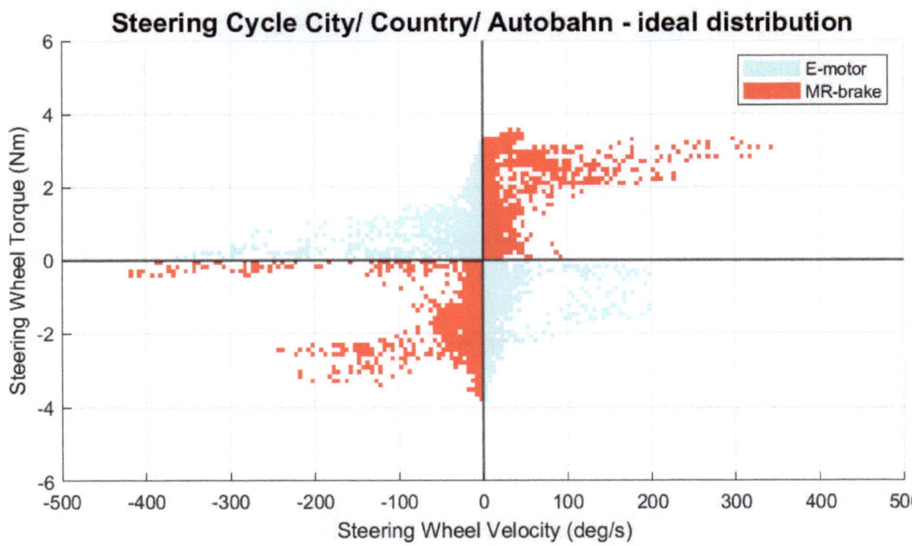

Fig. 5. E-motor and MR-brake ideal torque splitting for real test data

This representation does not provide a depiction of the occurrence frequency for each actuator. However, it is noteworthy that the instances of high steering wheel torque (SWT) and steering wheel velocity (SWV) occur less often.

To design a control system for distributing the torque with good steering feel, the control must be advanced. This includes blending of actuator torques in the steering off-center and more e-motor proportion in the on-center region, thereby moving away from the ideal areas. Sudden changes in e.g. SWV may lead to unwanted behaviors due to settling time and minor differences in actual torque.

To realize blending, factors between 0 and 1 were introduced and multiplied with the respectable torque command value of MR-brake and e-motor. These factors of the two command values behave inversely and are dependent on steering wheel torque and SWV, hence allowing areas where both actuators are actively setting torque.

The factors are realized via a characteristic field as shown in Figure 6. At lower SWVs, the MR-brake factor is 0 and reaches a factor of 1 at a given SWV, called transition speed. By using different transition speed values for low and high torques to blend, a finer adjustment can be made later. A full takeover of torque from the brake only happens in areas where the MR-brake factor is 1 (resulting in an e-motor factor of 0). In areas where the MR-brake is active, but the factor is smaller than 1, the e-motor is partially taking over the torque in low SWV section, achieving torque blending of each actuator torques.

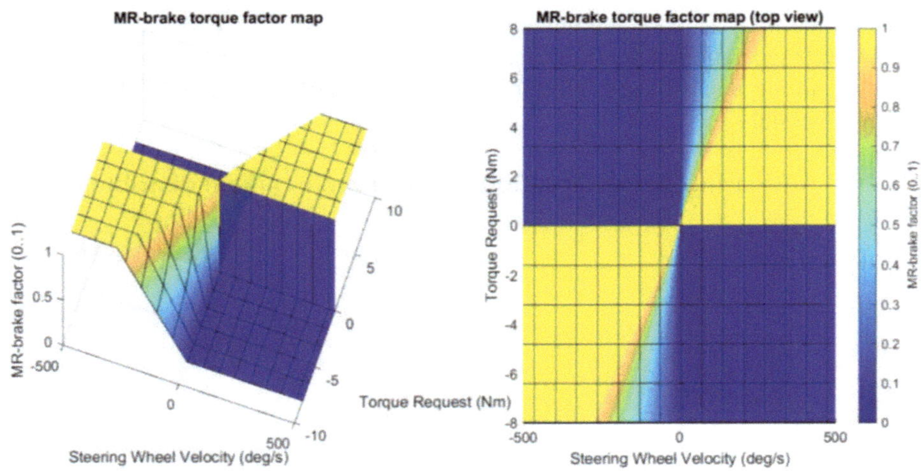

Fig. 6. Torque factor map of MR-brake for torque splitter

For end stop behavior, a similar logic is used with two main differences. Firstly, since the achieved SWVs are low compared to driving even when pushing over the end stop torque limits, the transition speeds of the characteristic field are kept low, resembling more the ideal torque splitter. Secondly, the e-motor is not disabled in areas where the MR-brake is active to avoid fast on and off switching of e-motor torque due to the steeper map. Steering into the end stops results in an added torque of e-motor and MR-brake, steering away from the end stops is only controlled by the e-motor.

Reevaluating the driving data but employing a simulation of output torque for both the MR-brake and the e-motor with the help of the new torque splitter, the diagram depicted in Figure 7 can be generated. The effect of blending is recognizable, equal to the torque factor map. Additionally, higher torque values have been distributed to MR-brake and e-motor, resulting in lower torques for each actuator in areas of active blending.

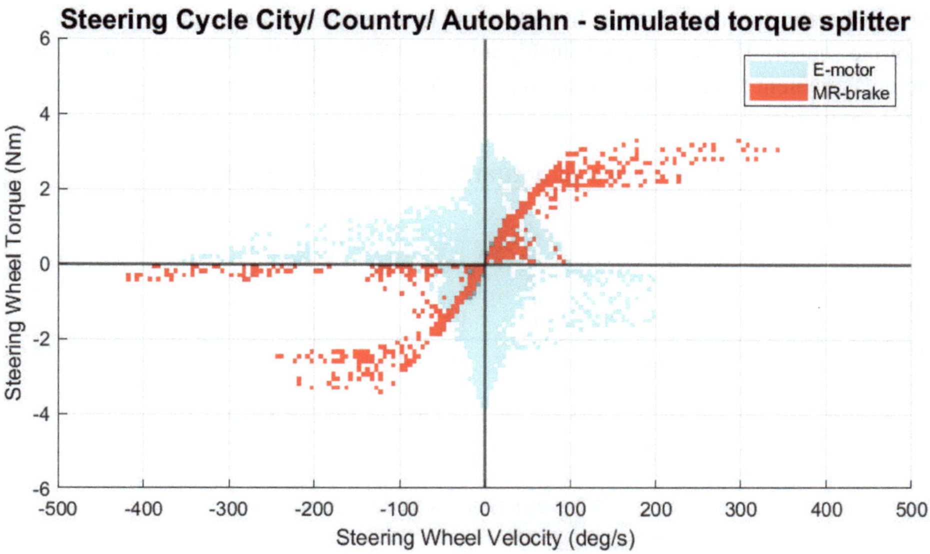

Fig. 7. E-motor and MR-brake simulated torque splitting for real test data

4 Simulation and Driver in the Loop Results

For the execution of Driver-in-the-Loop (DiL) tests, the hybrid actuator was implemented within a static driving simulator stationed in the laboratory of the Munich University of Applied Sciences. The control unit for the MR-brake and the inverter for the e-motor are connected to an embedded computer. This embedded computer handles the torque splitter and simulates a Pfeffer steering model [8]. The embedded computer is communicating with a real time capable computer via a client-server interface. Additionally, a vehicle simulation is running on the real time computer. Measuring data and controlling the simulation is done by a host computer connected to the real time computer.

To evaluate the functionality of the torque splitter, two different tests scenarios are performed.

1. Weave Test: Conducted at a velocity of 100 km/h, the test involved steering wheel angle (SWA) amplitudes approximating 30 degrees at a frequency of 0.2 Hz. This procedure achieved a peak lateral acceleration of 4 m/s².
2. End Stop Test: This test was performed at standstill, involving steering into one end stop.

For both scenarios the actual torque values were computed utilizing the currents and their respective torque constants. It's important to note that the setup did not incorporate a torque sensor. The upper graph in Figure 8 shows the torques of the actuators over the measurement time, in the lower graph the torques are plotted against the SWA for the Weave Test.

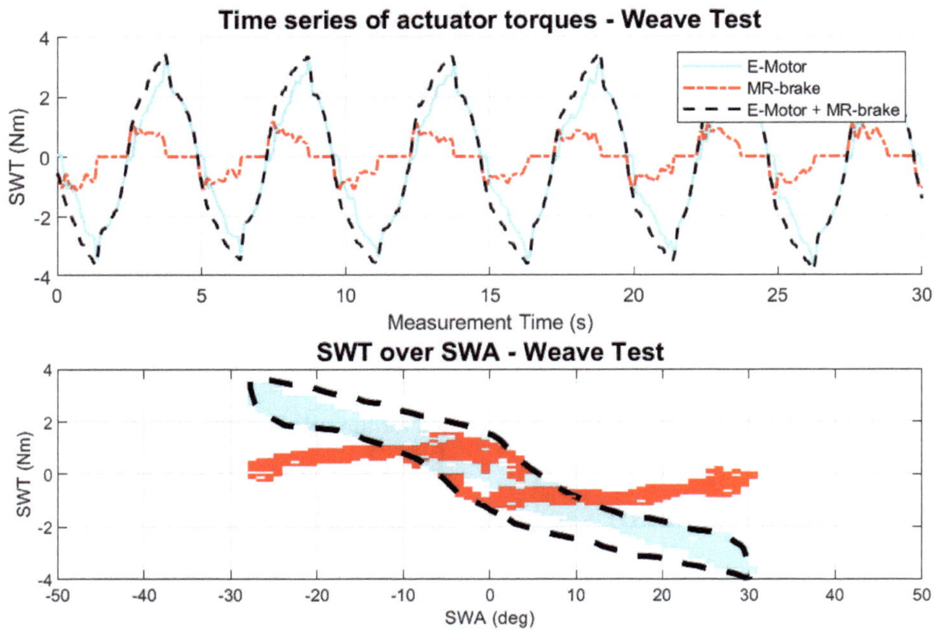

Fig. 8. E-motor and MR-brake torque splitting for Weave Test

The MR-brake becomes active during turning the steering wheel into the sine wave according to the torque-velocity diagram. Arriving at the peak of SWA sine wave, the MR-brake torque ramps down due to decreasing SWV. The e-motor takes over completely to generate the counter torque and a smooth transition can be seen. By superimposing the torques of each actuator, a well-defined hysteresis can be generated.

The upper graph in Figure 9 shows the torques of the actuators over the measurement time, in the lower graph the torques are plotted against the SWA for the End Stop Test. The MR-brake becomes already active during slowly turning the steering wheel towards the end stop and the e-motor is inactive. At 4.5 s, when the end stop limit is reached the MR-brake and the e-motor torques are rising to the parametrized maximum. Between 8 s and 10 s the end stop is slightly released and the MR-brake torque drops down to 0 Nm. At 11 s during steering back to the center the e-motor torque is reduced slowly as well. The end stop transitions can be generated as a sum of the actuators or independently as shown in the measurement.

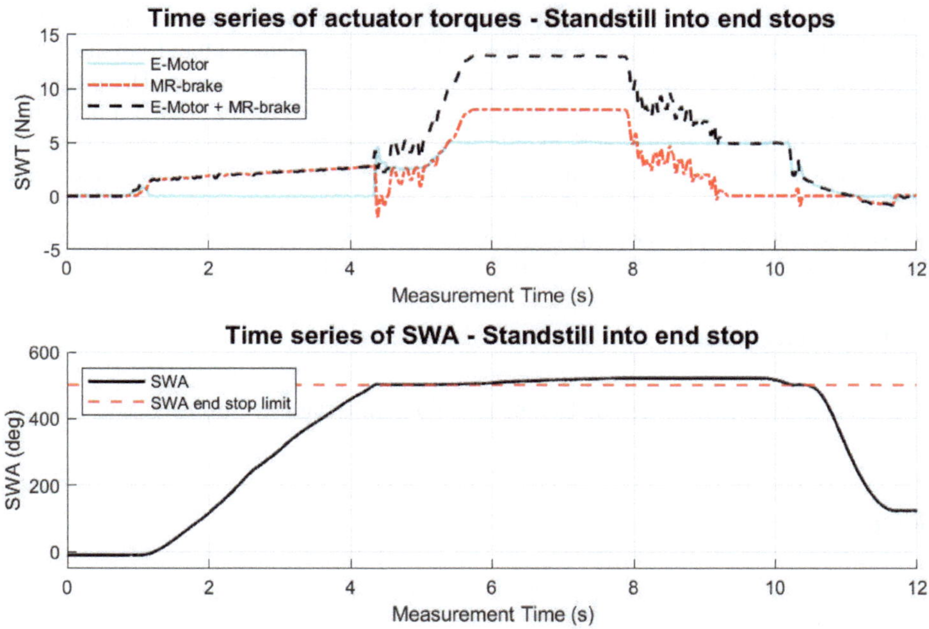

Fig. 9. E-motor and MR-brake torque splitting for End Stop Test

An expert evaluation of both scenarios in a DiL environment shows good subjective results, the transitions between the actuators where hardly noticeable, only the end stops where described as "little sticky". The MR-brake is not released early enough, because the driver's intention is sensed after oversteering the MR-brake torque. Changing the position of the angle sensor closer to the steering wheel will improve the sticky feel by using the systems elasticity similar to a torsion bar.

5 Summary and Outlook

In this study conventional steer by wire systems were analyzed and the potential advantages for a force feedback actuator with direct drive e-motor and MR-brake presented. The design is cost effective, delivers a high maximum torque at low energy consumption and ensures flexible end stop positions. The steering feel of the proposed design was analyzed objectively and evaluated subjectively by experts. The results show that the developed torque splitter feels already smooth in the transition between the actuators, but the MR-brake tends to stick in the end stops. The reason for the behavior was identified in the angle sensor position at the end of the actuators. The sensor type will be changed towards a hollow shaft type and positioned between the steering wheel and the e-motor to improve the sticky feel. The torque/angle input from the driver is measured early to release the MR-brake instead of first oversteer it. The proposed actuator design delivers a steering feel more equivalent to a direct drive e-motor only concept. A follow up study could include an artificial sensor quality reduction in terms of resolution and accuracy to identify the requirements for a series production system. A second demonstrator with a series production EPS-motor including an updated MR-brake is currently being built up to further investigate the potential of the presented force feedback actuator with direct drive e-motor and MR-brake.

Acknowledgement

This study was supported by the European Union "Interreg Bayern-Österreich".

References

[1] F. Wilhelm, T. Tamura, R. Fuchs and P. Müllhaupt, "Friction Compensation Control for Power Steering," in IEEE Transactions on Control Systems Technology, vol. 24, no. 4, pp. 1354-1367, July 2016

[2] https://www.thyssenkrupp.com/en/stories/automotive-and-new-mobility/steer-by-wire-thyssenkrupp-schlaegt-innovativen-weg-in-der-entwicklung-ein

[3] https://www.nexteer.com/columns-intermediate-shafts/stowable-steering-column/

[4] https://www.iwm.fraunhofer.de/en/services/manufacturing-processes/powder-technology-particle-simulation/magnetorheological-fluids.html

[5] https://www.inventus.at/de_de/rotary-haptic-solutions/steer-by-wire/

[6] https://simucube.com/simucube-2-pro/?v=3a52f3c22ed6

[7] Strecker Z, Roupec J, Mazurek I, Machacek O, Kubik M, Klapka M. Design of magnetorheological damper with short time response. Journal of Intelligent Material Systems and Structures. 2015;26(14):1951-1958

[8] Pfeffer, P. E., Harrer, M., & Johnston, D. N. (2008). Interaction of vehicle and steering system regarding on-centre handling. Vehicle System Dynamics, 46(5), 413-428.

Conception and Prototypical Implementation of an Integral Steering and Drive Module

Torben Dittmar[1] and Lutz Eckstein[1]

[1] Institute for Automotive Engineering (ika), RWTH Aachen University, Steinbachstraße 7, 52074 Aachen, Germany
torben.dittmar@ika.rwth-aachen.de

Abstract. The automation and electrification of vehicles facilitates the development of novel chassis systems. With the introduction of x-by-wire systems, the mechanical separation of actuators from control elements allows for very compact chassis modules that combine propulsion, steering and braking, also known as corner modules. Different variants of these modules are developed, but so far they employ all one actuator for driving and another, separate actuator for steering. A novel concept has been invented with the objective of optimizing system performance and size. This concept involves the assignment of actuators to multiple tasks, allowing them to be used for driving and steering simultaneously. In this concept, two electrical actuators are arranged coaxially with the steering axis and are connected to each other in the center of the wheel via a bevel gear. This implies that driving can be achieved through the equal torque of both actuators acting in opposite directions, while steering can be realized by torque acting in the same direction. Furthermore, both load cases can also be superimposed. The maximum power of both actuators can thus be split up as desired between steering and driving. Therefore, the size of the actuators can be adapted to the maximum required system power. However, since the overall system behavior is not stable, it is essential to design a dedicated control system that is robust and reliable. This paper presents the theory and control algorithm of the novel corner module concept and offers the initial promising results of a 48 V powered prototype.

Keywords: corner module, steering system, by-wire, wheel hub drive

1 Introduction

The transformation of the automotive industry and its main trends, including electrification, digitalization and modularization, lead to significant changes in nearly every subsystem of the entire vehicle and drive further technology developments.

The electrification of the drive system means that hydraulic components are increasingly being replaced by electric actuators even in addition to the components originally driven by the combustion engine. Examples of this include steering and brake actuators. In contrast, digitalization is leading to a shift in the design paradigm for vehicle development from mechanics to software. However, consistent software implementation in combination with electrification requires an adapted E/E architecture for the vehicle

P. E. Pfeffer (Ed.), *15th International Munich Chassis Symposium 2024*,
Proceedings, https://doi.org/10.1007/978-3-662-71064-7_10

that enables the integration of new subsystems in a modular and uncomplicated way on basis of a central ECU. One objective is to create new functions based on existing actuators and make them available to the user via an update even at a later stage of the vehicle's lifecycle. However, automated driving is also dependent on these development trends, as both sensors and actuators of the vehicle must be able to be monitored and controlled centrally by an ECU independently of the driver. [1]

Besides, modularization results in a simple basic structure of the electrified vehicle, which is often referred to as a skateboard chassis. This platform contains all essential elements for driving, including power supply. Based on this, only a specific hat for the brand and intended use for the customer needs to be added on top. Thanks to digitalization and electrification the hat and the platform can be easily linked together logically. The necessity for complex mechanical connections, such as for the steering system, is eliminated. This, in conjunction with the implementation of automation, necessitates the use of by-wire systems as illustrated in Fig. 1. Additionally, novel HMI (human-machine interface) concepts such as sidesticks can also be implemented in the vehicle as an alternative to the steering wheel based on by-wire technology [2].

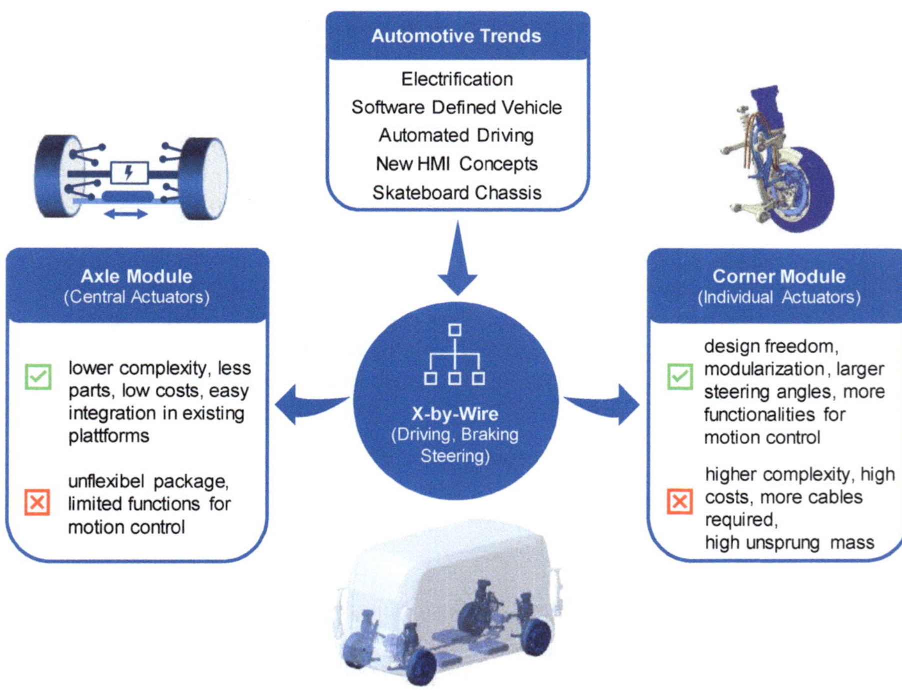

Fig. 1. Automotive trends lead to X-by-Wire development and thus to new possibilities for the subsystem chassis

With regard to the chassis subsystem, the prevailing development trends result in electrified and logically controllable chassis modules. These can be designed in either a conventional manner as a whole axle module or as single wheel corner modules that contain the basic functions of driving, braking, steering and suspension. The use of an axle module allows the centralization of some actuators, thereby reducing both complexity and costs. However, the use of two or four singe wheel corner modules on a vehicle offers more degrees of freedom for motion control, as each actuator can be controlled individually. In addition, the required package can be defined more freely, as it is not linked to other components of the car. [1]

A variety of variants of these corner modules have already been developed and numerous publications have been written on this topic, e.g. [3]–[11]. However, these variants employ a single actuator for driving and a separate, distinct actuator for steering. One example is the corner module, which was developed for automated vehicles in the public project UNICARagil, which was funded by the German Federal Ministry of Education and Research (BMBF) [12]. When designing these actuators, it is essential to consider the maximum load case for each of them. This specifies the required actuator size. Upon comparison of the different load cases, it becomes evident that the maximum power requirement for steering is at standstill. Conversely, for the propulsion system, the maximum power is needed at higher speeds. Therefore, the power required at the same time is less than the total power of the overall system. This leads to larger actuators than necessary, resulting in increased weight and installation space. Moreover, the routing of cables and cooling hoses for corner modules with high steering angles of up to 90° represents a significant challenge. This is due to the necessity of guiding theses flexible components and the fact that they are exposed to high loads due to the large rotational movements.

In order to optimize system performance and size, a concept for an integral steering and drive module has now been developed. This concept involves the use of actuators that are no longer exclusively assigned to one task. Instead, they can be used for driving and steering simultaneously. In this concept, two actuators are arranged coaxially to the steering axis and are connected to each other in the center of the wheel via a bevel gear. This implies that driving can be achieved by the combined torque of both actuators acting in opposite directions, while steering can be realized by the torque acting in the same direction. Furthermore, both load cases can be superimposed. Thus, the maximum power of both actuators can be distributed as desired between steering and driving. Therefore, the size of the actuators can be adapted to the maximum required system power. Besides, an optimized arrangement of the electric motor facilitates improved cable routing, eliminating rotational movement of the power and logical cables. As the overall system behavior is inherently instable, it is essential to implement a robust control system.

This paper introduces the concept, theory and control algorithm of this integral steering and drive module, followed by the implementation of a first prototype, that is designed for L7e vehicles. Based on this, the first test rig results are presented and discussed. The last section concludes the information presented in this paper and gives an outlook of the next steps on this research topic.

2 Concept overview

This chapter presents the concept of the new integral steering and drive module. First, the requirements for the development are identified and then the underlying mechanical concept is explained. Finally, various scenarios for the superimposition of steering and driving are presented as examples.

2.1 Requirements

The Institute of Automotive Engineering at RWTH Aachen University has been engaged in research on steer-by-wire systems and corner modules for several years. The SpeedE research vehicle features a wheel-individual steer-by-wire system in which an electric steering actuator installed above the wheel carrier can steer the wheel from -60° to 90° [13]. In the UNICARagil project, this system was supplemented by an electric 48 V wheel hub motor. The system achieves a steering angle of -66° to 93° [12]. In both vehicles, the wheel brake can also be controlled electrohydraulically by wire. Both systems have in common that the electric motors for the steering actuators are larger than the power of normal electric steering gear actuators. In the UNICARagil vehicles, these electric motors have a rated output of 3.6 kW. This corresponds to approximately one-third of the power of the traction motor. This power is required due to the large bore torque that occur when the wheel is turned around the steering axis while the vehicle is not in motion. In addition to the tire and the wheel load, the size of the scrub radius has a decisive influence on the size of the bore torque. The maximum steering torque becomes smaller as soon as the wheel rolls. At higher lateral acceleration and rising slip angle, however, a steering torque requirement of approximately one-third of the bore torque in stationary situation arises. This means that two-thirds of the power of the electric motor is only required for one operating point, steering for parking when stationary [12], [14]. Cases of misuse, such as a heavy approach to a kerb, were not taken into account here.

Due to their dimensions, the steering angle actuators could not be compactly integrated into the wheel suspension. Among other things, this led to a larger scrub radius which also results in a larger bore torque and thus steering torque. The additional installation space is located above the universal joint of the corner module and was disadvantageous for the development of the interior of the UNICARagil vehicles. Another challenge was the cable routing for the 48 V wheel hub motor. It had to be ensured that the cables could permanently withstand both a rotation of up to 90° and a simultaneous vertical movement of ±80 mm without colliding with other components.

This results in the following requirements for the new corner module concept:

- Implementation of drive, steer and brake-by-wire
- Compact integration of the steering angle actuator
- Realization of large steering angles up to 90°
- Minimization of the scrub radius
- Simple cable routing without extensive movements
- Compact package of the total module and partial integration into the wheel

- Possibility for a simple suspension and module mount

The integration of the suspension was not formulated as a core requirement since the suspension should be selectable depending on the vehicle concept.

2.2 Mechanical concept

An invention by Lutz Eckstein was selected as the framework for the novel corner module concept to meet the aforementioned requirements [15]. This concept employs two actuators, both of which can be utilized to accelerate and decelerate the wheel as well as to adjust the steering angle. This has the advantage that the total power of the installed actuators can be minimized. A schematic illustration of the concept can be found in Fig. 2.

In this corner module, two coaxially arranged actuators are connected in the middle to a bevel gear drive, on whose output the wheel hub is mounted. The axis of rotation of the two actuators is also the steering axis. Ideally, the entire module is arranged in the middle of the wheel, which minimizes or eliminates the scrub radius. An equal torque of both electric motors M_{UE} and M_{LE} causes a torque $M_{Steering}$ around the steering axle, while an opposite torque of equal magnitude causes a torque M_{Drive} around the drive axle. It is also possible to superimpose both cases to enable simultaneous driving and steering. The resulting moments of the actuators are supported by a suitable wheel suspension. Depending on the wheel suspension selected, this can be realized by a single fixed connection or separate links for each actuator side.

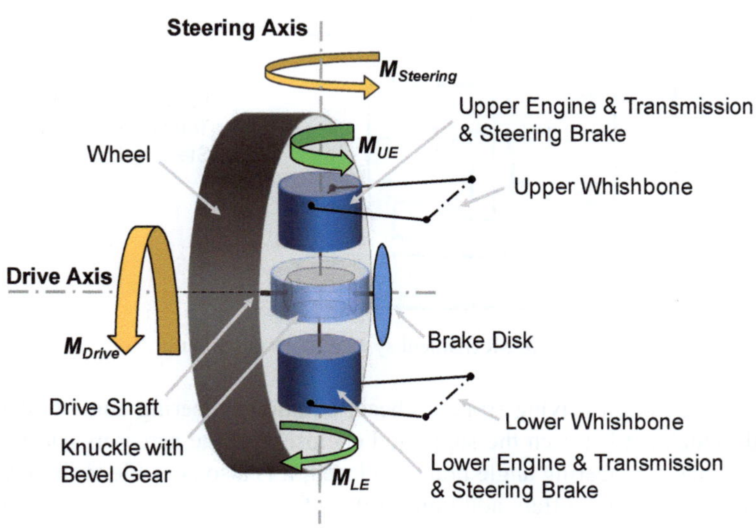

Fig. 2. Schematic drawing of integral steering and drive module

Depending on the package requirements, the wheel brake can be designed on the side of the wheel output or on the opposite side, using an additional bevel gear. In the latter case, the brake disk rotates in the opposite direction to the wheel.

In order to optimize the performance of the electric motors, which ideally rotate at high speed, it is reasonable to integrate an additional transmission between the actuators and the bevel gear. In the illustration of the overall mechanical system in Fig. 3, the intermediate transmission is indicated by a planetary gearbox. However, other types of transmissions can also be used. Thus, the ratio i_{St} converts the torques and speeds of the actuators to the necessary steering torques and steering angle speed. The drive torque and the wheel speed are subsequently determined by the transmission ratio of the bevel gear i_{BG}.

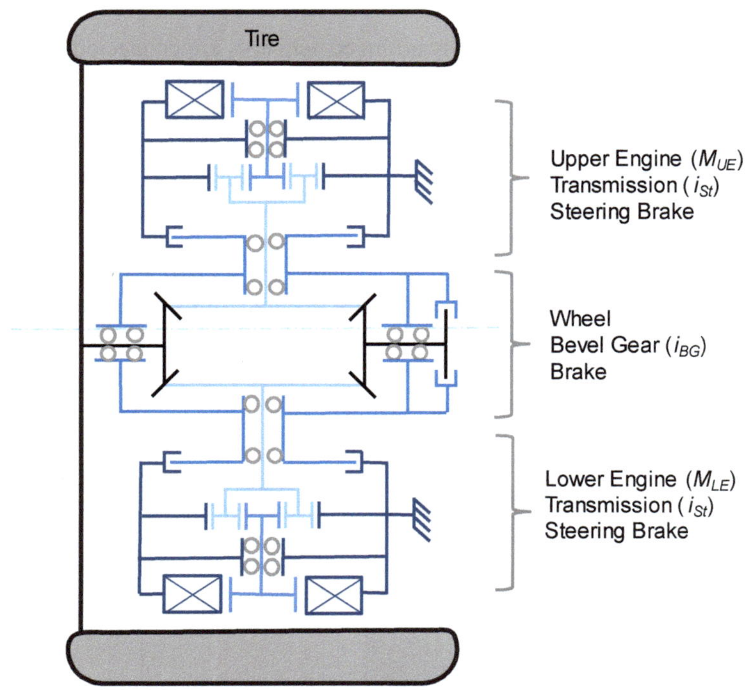

Fig. 3. Mechanical system of corner nodule

To save energy when driving straight ahead, additional steering angle brakes can be installed, which act between the supported enclosure of the actuators and the steered knuckle. When the steering angle brake is closed, it is also possible to drive with only one actuator. This also offers additional safety functions in the event of a technical defect in one of the actuators. A sophisticated combination of using wheel brake and steering angle brake simultaneously, the steering angle can be influenced and brought into a tolerable range, at least for a short time, while driving, in addition to maintaining the drive torque.

The maximum steering angle achievable is contingent upon the geometric and kinematic boundary conditions. The selected scrub radius, kingpin angle, rim width, and inside diameter, in addition to the dimensions and attachment of the control arms to the corner module, exert a considerable influence on this outcome. With optimally selected characteristic values, a steering angle of ±90° can be attained. However, this is also associated with disadvantages in the distribution of the applied forces and the controllability, so that a suitable compromise of these characteristic values must be found for the respective vehicle concept.

2.3 Scenarios for steering and driving

According to the mechanical system, the equations for the steering and drive torque, without consideration of friction, are as follows:

$$M_{Steering} = i_{St} \cdot (M_{UE} + M_{LE}) \quad (1)$$

$$M_{Drive} = i_{BG}\, i_{St} \cdot (M_{UE} - M_{LE}) \quad (2)$$

The drive and steering power depend on the speed of movement of the respective subsystem and are calculated according to the following equations, without taking efficiency into account:

$$P_{Steering} = M_{Steering} \cdot \omega_{Knuckle} = (M_{UE} + M_{LE}) \cdot i_{St} \cdot (\frac{\omega_{UE} + \omega_{LE}}{2}) \quad (3)$$

$$P_{Drive} = M_{Drive} \cdot \omega_{Wheel} = (M_{UE} - M_{LE}) \cdot i_{BG} \cdot i_{St} \cdot (\frac{\omega_{UE} - \omega_{LE}}{2}) \quad (4)$$

The total power of the corner module can be determined by either summing the power of the driving and steering outputs or by summing the powers of both electric motors. In this context, losses are not considered.

$$P_{Overall} = P_{Steering} + P_{Drive} = M_{UE} \cdot \omega_{UE} + M_{LE} \cdot \omega_{LE} \quad (5)$$

Upon examination of the various scenarios for the combination of drive and steering on the aforementioned basis, as illustrated in Fig. 4, it becomes evident that the maximum power cannot be consistently achieved in all cases due to the underlying concept.

In scenarios 1 and 2, the maximum torque of both electric motors is employed for pure driving or steering. This allows the maximum power of both actuators to be used. With a torque split of 50% each for driving and steering, the torque from driving and steering on one actuator cancels each other out. This is demonstrated in scenario 3. A closer look at the formula for total power reveals that only half of the available system power is accessible in this instance. However, should a higher power output be required, it is possible to reduce the drive torque, which not only enhances the available torque for steering but also increases the maximum total system power. This phenomenon can be observed in Scenarios 4 and 5. It is crucial to consider this fundamental fact when developing the control system for the corner module.

	1	2	3	4	5
Diagram	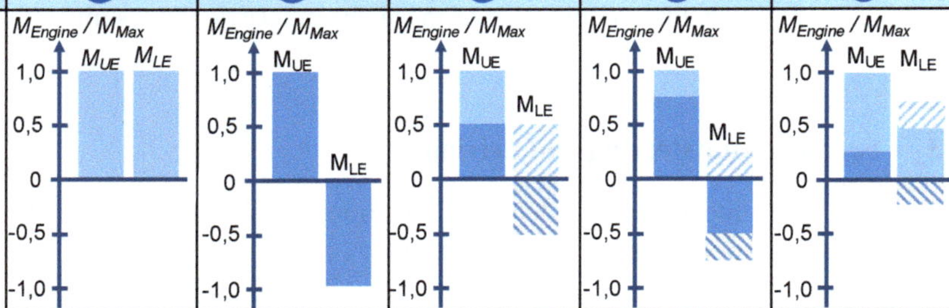				
Scenario	max steering only	max driving only	50 % steering 50 % driving	75 % steering 25 % driving	25 % steering 75 % driving
Steering Torque	$2 * i_{St} * M_{max}$	0	$i_{St} * M_{max}$	$0,5 * i_{St} * M_{max}$	$1,5 * i_{St} * M_{max}$
Drive Torque	0	$2 * i_{St} * i_{BG} * M_{max}$	$i_{St} * i_{BG} * M_{max}$	$1,5 * i_{St} * i_{BG} * M_{max}$	$0,5 * i_{St} * i_{BG} * M_{max}$
Max Overall Power	$2 * P_{Engine\_max}$	$2 * P_{Engine\_max}$	P_{Engine_max}	$1,5 * P_{Engine\_max}$	$1,5 * P_{Engine\_max}$

torque for driving — toqure for steering — neutralized torque for driving — neutralized torque for steering

Fig. 4. Scenarios of combining driving and steering with max overall power

3 Control strategy

The distinctive feature of the control strategy of this corner module concept is the challenge of regulating two independent control variables, namely the steering angle $\varphi_{Steering}$ and the wheel speed n_{Wheel}, through the torques M_{UE} and M_{LE} of two coupled actuators. Instead of the wheel speed, which is proportional to the vehicle speed $v_{Vehicle}$, a drive demand torque T_{Drive} can also be selected as the second control variable. To address this challenge, the system is initially divided into two individual control loops. Cascade controllers serve as the basis for both. As illustrated in Fig. 5, the steering angle difference between the target and actual values is initially used to determine a steering speed using a P controller. This steering speed is then compared to the actual steering speed and used as an input variable for a PI controller to calculate the torque for the actuators. The maximum steering angle speed can be limited in the P controller. The torque required for adjusting the steering angle is divided equally between the two actuators.

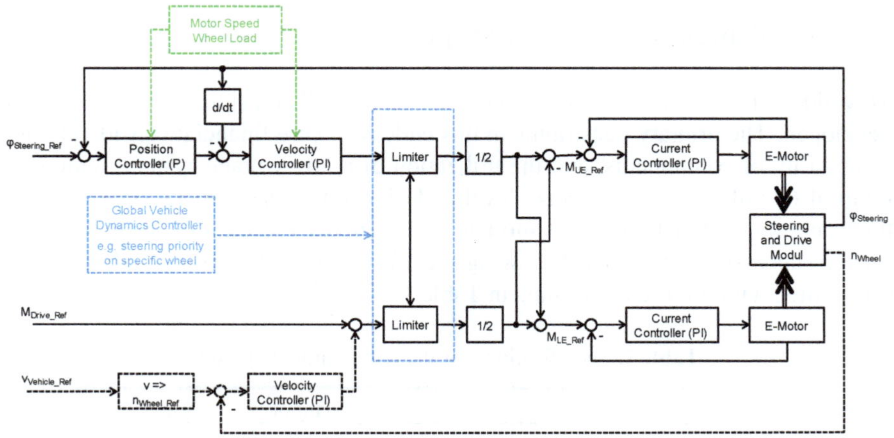

Fig. 5. Control concept for steering and drive module

If we now consider the second control path for the wheel speed, this behaves in the same way, but no position control is required. The required actuating torque can be determined directly from the comparison of the target and actual wheel speed using a PI controller. According to the equations in chapter 2.3, half of the torque is passed on to the two actuators with different signs. The resulting torque is employed as a reference variable for separate current control with a PI controller for each actuator.

In this context, it is possible to consider the two control loops separately from each other. However, as the maximum available torque and the power of each actuator are finite, it is necessary to limit the required torque before distributing it to the actuators. This must be determined by a global vehicle motion controller depending on the driving situation. In general, steering requires smaller torques than driving torques and should be prioritized in most cases for safety reasons. As illustrated in Chapter 2.3, a specific reduction of the drive torque can also lead to an increase in overall performance, conferring advantages in certain driving situations. Special cases, such as driving over a kerb while parking, must be considered separately.

For the stability and performance of the control system, it is also important to consider the engine speed and wheel load, especially for the positioning of the steering angle. When the vehicle is stationary, the actuators are also at standstill. In this operating state, it is more challenging to precisely adjust the angle, depending on the selected transmission ratio and the resolution of the encoders of the electric motors. The elasticity and mechanical play within the system also play a pivotal role. If the vehicle is moving, the actuators also have a certain speed. The existing rotary movement precludes any potential for play in the system. Adjusting the steering angle by applying an additional torque is easier while the electric motor is rotating. It is also worth noting that a larger transmission ratio generally eases the control of the steering angle. However, this is at the expense of the maximum vehicle speed, depending on the bevel gear ratio. Therefore, an optimal compromise between the two ratios must be found for the required steering and drive torque, the maximum vehicle speed, and the optimal control.

4 Implementation of prototype

To validate the promising simulation outcomes, a small and simple prototype has been developed. The primary constraints in this endeavor were financial considerations, the availability of components, and the utilization of a low-voltage system, with a maximum of 60 volts. The low voltage level is designed to ensure operational safety. Furthermore, the installation of this module in a compact vehicle should be feasible following successful testing. The following specifications for the corner module have been established within these constraints in Table 1:

Table 1. Specifications for the corner module prototype

Parameter	Symbol	Value
Vehicle category		L7e
Voltage of the electric motors	U_{Supply}	$\leq$ 48 V
Continuous power per module	P_{Module_Max}	2 kW
Maximum vehicle speed	$v_{Vehicle\_Max}$	50 km/h
Maximum continuous steering torque	$M_{Steering\_Max}$	$\geq$ 100 Nm
Maximum continuous drive torque	M_{Drive_Max}	$\geq$ 30 Nm

The specified torques represent the maximum torque that can be provided in a permanent manner. The peak torques can be considerably higher. Prior to the selection of gear ratios, a suitable motor is identified. This motor is characterized by the date in Table 2:

Table 2. Specifications of the selected electric motor

Parameter	Symbol	Value
Supply voltage	U_{Supply}	48 V
Rated torque	T_{Rated}	1.76 Nm
Peak torque	T_{Peak}	4.95 Nm
Rated current	I_{Rated}	16.9 A
Peak current	I_{Peak}	47.6 A
Maximum speed at rated torque	n_{Rated_Max}	3230 rpm
Maximum speed	n_{Max}	4800 rpm

The data indicates that the gear ratios can be set to $i_{St} = 30$ and $i_{BG} = 1/3$. The calculated maximum continuous steering torque is therefore $M_{Steering\_Max} = 105.6$ Nm, the maximum continuous drive torque $M_{Drive\_Max} = 35.2$ Nm and the maximum possible vehicle speed, taking into account a dynamic tire rolling radius of $r_{dyn} = 265$ mm at $v_{Vehicle} = 48$ km/h. This indicates that the speed is minimally below the required boundary conditions.

The overall structure of the prototype is shown in Fig. 6. A double wishbone axle with a spring/damper unit acting on the lower wishbone is selected as the wheel suspension. The high angle of the spring is due to a simple installation and the limited package, and thus does not play a significant role in the investigations of the corner module. The kingpin angle, caster angle, as well as the offsets to the steering axle and the wheel contact patch are set to zero. The scrub radius is 48 mm. It is not possible to further reduce the scrub radius due to the given inner shape of the rim and the height of the overall module. However, it would be possible to reduce the scrub radius to zero with an adapted rim. The geometric configuration, in conjunction with the rim, yields a steering angle of ±55°.

Fig. 6. Prototype of steering and drive module with double wishbone suspension

The electric motor and gearbox are located in the outer parts of the module housing. Due to the large transmission ratio and for packaging reasons, a strain wave gearing is used instead of a multi-stage planetary gearbox. The 24V-operated steering angle brakes are located underneath on the inside, and a 20-bit steering angle encoder is located on both sides of the module.

The white enclosure of the bevel gearbox represents the steering knuckle, to which the speed sensor and the brake caliper are also attached, can be seen in the center of the module. For cost reasons, a off-the-shell component is used, which is why no second output shaft is available and therefore the brake cannot be mounted to the inside as shown in Fig. 3. Due to the existing connections of this bevel gearbox and the lack of integration options in the rest of the design, the package is longer than necessary. With an integral overall design, a further reduction of the overall height can be assumed. The dimension of the wheel brake system, which in this case is actuated by a cable pull, is relatively small. This is also due to the restrictions of low-budget components, that are compatible with the package. However, the performance of the brake is sufficient for test bench operations at given wheel speeds.

All electrical components are arranged in such a way that, with the exception of the speed sensor and the brake, all other cable connections are fixed to the drive housing and do not rotate with the knuckle. This configuration allows for simple cable routing via the wishbones, and the cables only have to withstand a small movement in the transition between the wishbones and drive pots during vertical wheel travel. With the exception of a few components within the drive unit, the control arms and all housing parts are manufactured from aluminum.

With the exception of the power supply, all the electronics are combined in a compact layout, as can be seen in Fig. 7. Two TI C2000 microcontroller experimental boards serve as the basis. The first microcontroller board contains two 3-phase gate drivers for controlling the electric motors. As these are only limited to a voltage supply of 45V, the voltage was initially reduced to 36V for the inverters of the drives. The combination of both inverters on a single board offers a significant advantage, as it eliminates the potential for communication delays between the control system and both inverters. This eliminates a potential source of error when examining the performance of the corner module.

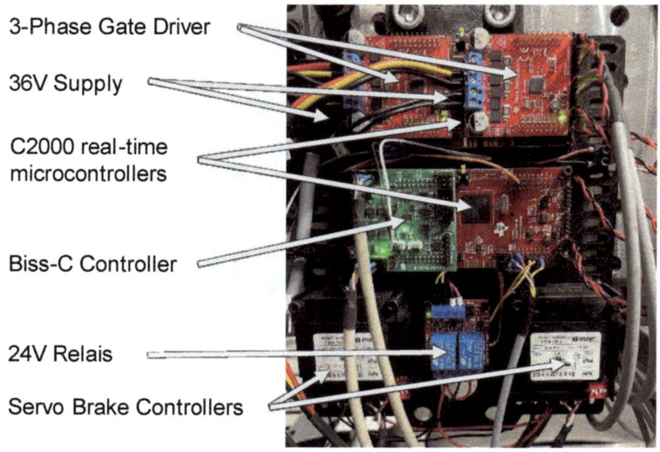

Fig. 7. Electronic prototype board including gate drivers for the electric motors

A second microcontroller board is used for the Biss-C controller, which is required for communication with the steering angle encoders. This pad also controls the cable actuator of the wheel brake and the steering angle brakes. Both launch pads communicate with each other via a CAN bus. In conjunction with the servo brake controllers, the 24V relay controls the steering angle brakes. All necessary sensors are supplied with a voltage of 5 V. The underlying controllers, including the control of the inverters, were created using Matlab/Simulink software and flashed to the two launch pads.

To assess the performance of the corner module, it is installed on the in-house NVH test bench developed by ika. This test rig facilitates the analysis of the vibration properties of complete suspension systems. Furthermore, it is employed for the validation of suspension CAE models, which support the development of NVH target settings on a system level under controllable boundary conditions. One notable feature of the chassis NVH rig is its ability to change the excitation type between the hydraulic setup and the drum setup. In the hydraulic setup, the excitation is generated by two hydraulic cylinders connected to the knuckle via an in-house developed wheel substitute system. The hydraulic excitation is capable of reaching up to 250 Hz and is designed to simulate the excitation of an asphalt surface [16]. However, for the purposes of this research, only the drum setup will be utilized. The suspension of the corner module is mounted on the axle frame, which can be moved in the vertical direction by two hydraulic cylinders. The configuration is illustrated in Fig. 8.

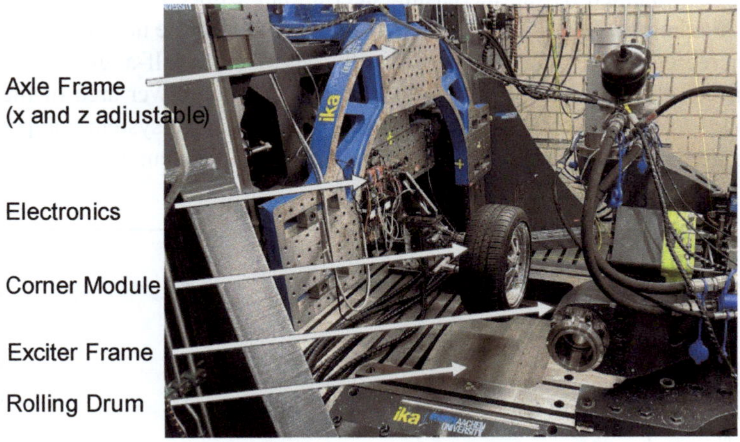

Fig. 8. Corner module prototype on ika NVH test rig

To investigate the performance and stability of the integral steering and drive module, both the stationary wheel, the self-driven wheel, and the wheel driven by the test bench drum are examined. The inertia of the test bench drum is equivalent to that of a compact car. Consequently, only a minimal acceleration is possible when using the corner module's own drive, and it must be supported by the drive of the test bench drum in certain scenarios.

5 Test rig results

The following results provide an initial answer to the general feasibility of the corner module concept described above. The aim of the investigations is to determine the stability of the inherently unstable system, the speed and accuracy of the steering angle position and the influence of the mutual influence of steering and drive torques. The basic controller application required for this is only based on a rudimentary setup with a focus on stability and does not represent the limit of the possibilities. In particular, the current controllers of the motors, some of which were designed in-house, offer further potential for improvement. However, this is not the focus of this study and will not be discussed further in this paper.

Fig. 9 depicts the hysteresis curve of a sinusoidal excitation of the wheel steering angle with a frequency of 0.2 Hz and an amplitude of ±40° while stationary for two different wheel loads. The direction of the curve is clockwise. As anticipated, a higher torque is required for an increased wheel load. Upon approaching the edge positions of *40°*, an increased oscillation can be observed. This phenomenon can be attributed to the flattening of the sinusoidal curve of the steering angle speed and the resulting slower rotational speed of the electric motors. As only Hall sensors are installed in the actuators to detect the position and speed, the current control is inaccurate at slow speeds. In contrast, a rounded curve can be observed when extending from the edge positions. This can be explained by the increasing steering angle speed of the sinusoidal curve until the maximum wheel steering torque is built up. Due to the neutral wheel kinematics, no increased torques can be observed at the edges, as no self-aligning torque builds up that needs to be overcome. The vibration curve in the lower area of the diagram corresponds to half a motor revolution. However, the control system compensates for this, so that the sinusoidal speed curve of the steering movement is not affected.

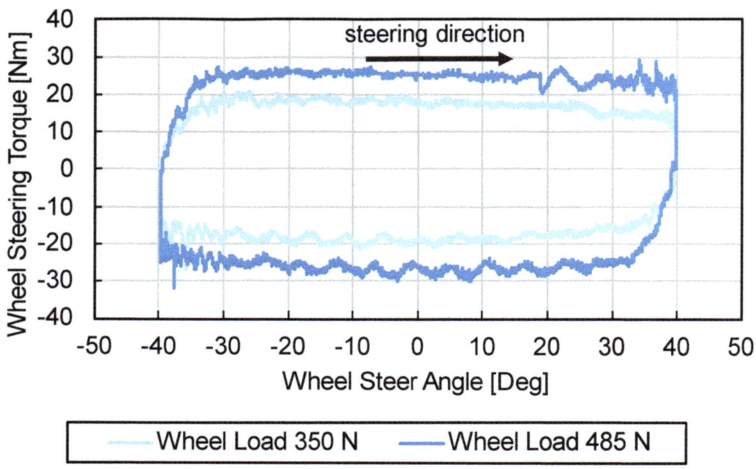

Fig. 9. Torque hysterese of full steering at standstill under different wheel loads

A Bode plot was created by carrying out sine sweep steering angle excitations with an amplitude of 1° and a frequency of up to 10 Hz for different wheel speeds. The result is shown in Fig. 10. It can be seen immediately that the control can only be carried out inadequately when the wheel is stationary at slow and small steering movements due to the sensors installed in the actuators. The magnitude of the Bode plot varies even at low frequencies and drops significantly at 1 Hz. Additionally, a phase distortion is evident at the outset. However, at higher wheel speeds, the curves are approximately the same. Up to a frequency of 3 Hz, magnitude and phase remain at low values. Only from this frequency does the phase drop, and from 8 Hz the magnitude also drops. According to the Nyquist criterion, the control remains stable in all cases. This suggests the potential for developing a faster control algorithm.

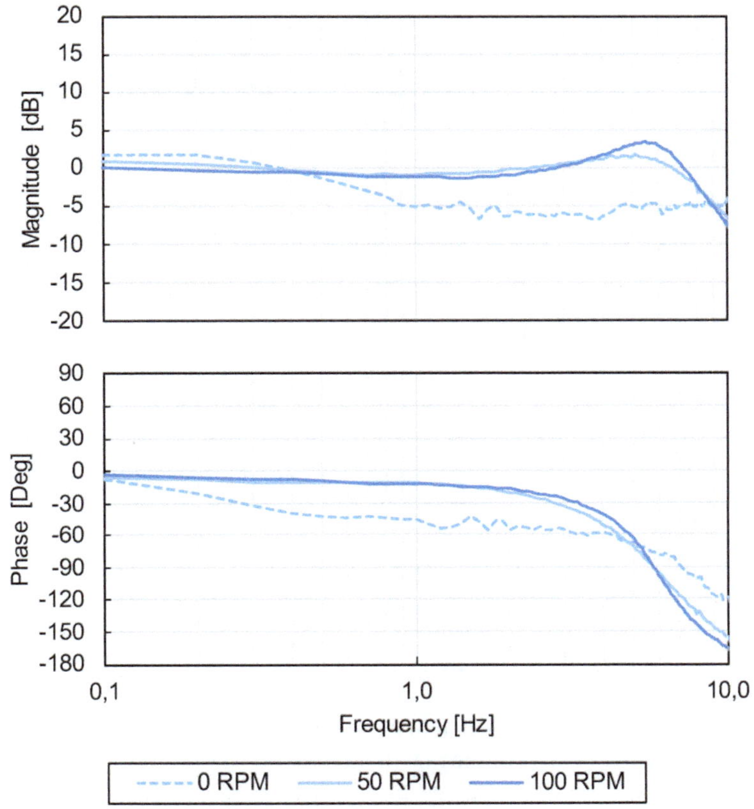

Fig. 10. Bode plot for 1° amplitude under standstill and rolling condition

The step steer is suitable for analyzing the controller's response time and speed control. Fig. 11 shows step steers of 1°, 5° and 10° both from a standstill and at a driven wheel speed of 60 rpm. When the wheel is stationary, the controller's response is conservative, and the steering angle reference is not set until after 0.6 s. This demonstrates that the controller is less responsive and less accurate in this condition. It is therefore necessary to consider a separate issue in the controller in any case. In addition, the use of higher-resolution encoders in the actuators must be considered. In contrast, the control system with a driven wheel is significantly faster and sets the desired angle consistently after 0.3 seconds. However, a clear overshoot can be seen at the beginning. It can be seen that the basic application is optimized for a step steer of one degree. The angle is reached after 0.1 s and remains stable from 0.2 s after the step steer. The vibrations observed from 0.6 s onwards at a step steer of 10° are due to the high slip angle and the resulting vibrations in the tire. The maximum steering angle speed of the wheel is 83 °/s. In a conventionally steered vehicle, this corresponds to a steering wheel speed of 1245 °/s with an assumed wheel to steering wheel ratio of $i_{SteeringWheel} = 15$.

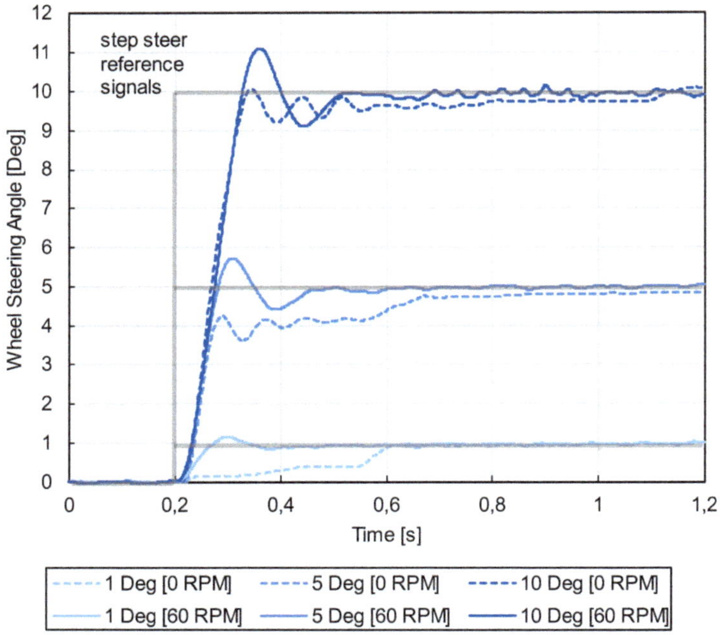

Fig. 11. Steep steer for 1°, 5° and 10° wheel steering angle

Fig. 12 illustrates the superimposition of a sinusoidal steering actuation with a constant specified drive torque of 25% of the maximum torque.

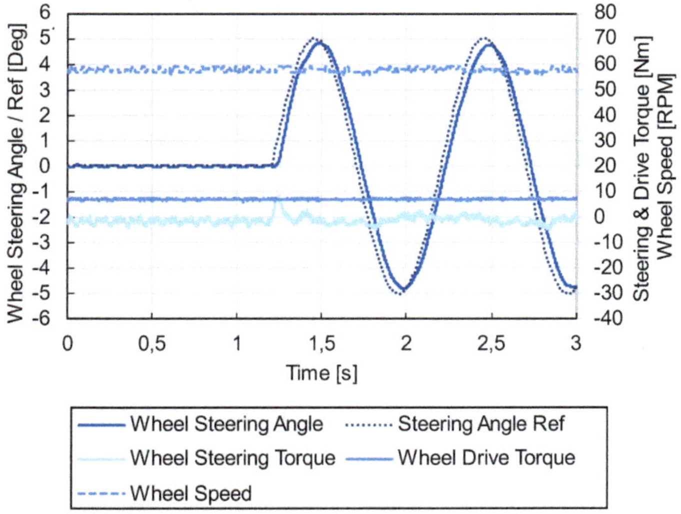

Fig. 12. Sinus steering while driving with constant torque

It is obvious that the steering angle lags behind the target value by approximately 0.05 s, and the amplitude is therefore not quite reached by approximately 0.2°. The steering torque required for this can also be regarded as sinusoidal, with the exception of a peak at the beginning. However, despite the sinusoidal torque of the steering, the specified drive torque and wheel speed remain almost constant. Only the braking effect of the slip angle of 5° results in a slight change in wheel speed. It can therefore be concluded that there is no visible mutual influence between the two control loops.

In comparison to the previous diagram, Fig. 13 illustrates the influence of varying longitudinal forces on the steering. To this purpose, the vehicle is accelerated from a starting speed of 1 km/h to 15 km/h within 2.6 s. It can be observed that the steering angle oscillates more strongly with the initiation of the acceleration process, and that the amplitude of the peaks increases from approximately ±0.03° to ±0.05°. Nevertheless, it can be stated that the steering angle fluctuations are generally at a very low level. Due to the time required for the tire to build up lateral force, the steering angle oscillations shown will not result in any significant lateral movement of the vehicle. The extent to which these vibrations influence tire wear must be investigated separately. When examining this diagram, it should be noted that the scrub radius of 48 mm exerts a significant influence on the forces acting around the steering axle.

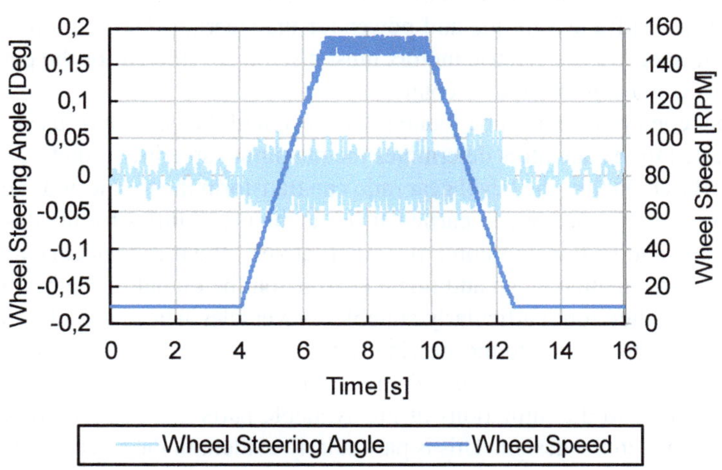

Fig. 13. Controlling constant steering angle under acceleration and braking

6 Conclusion and outlook

A novel concept and initial prototype for an integral steering and drive module were presented in this paper, and the initial results indicate the feasibility of this corner module concept. Additionally, fundamental stability of control has been demonstrated.

The primary objective of this corner module is to optimize system performance and size. This is achieved by utilizing two actuators that are no longer exclusively assigned to one task, but can both be used for driving and steering. In this concept, two actuators are arranged coaxially to the steering axis and are connected to each other in the center of the wheel via a bevel gear. This configuration enables the realization of driving and steering functions through the application of opposing torques from both actuators. The resulting compact package of steering and drive actuators allows for the integration of the entire module into the rim. According to the theoretical framework of this concept, steering angles up to 90° are achievable in an optimized configuration, depending on the rim, the control arms and the scrub radius.

Another advantage of this system is the simplicity of cable and cooling hoses routing. The arrangement of components presented in this paper allows for the routing of all cables, with the exception of those for the speed sensor and brake, in a manner that accounts for only wheel travel movement. This is primarily due to the positioning of the stator of the electric motor on the fixed side of the knuckle.

Upon consideration of various scenarios for combined steering and propulsion, it becomes evident that the maximum power of both electric motors cannot be fully utilized in all cases. In scenarios involving pure steering or driving, 100% of the power of both electric motors can be utilized. However, in the most unfavorable case of an equal distribution of the torques of 50% to drive and steering, only half of the total power can be used. In the control strategy, this situation must be taken into account by a global dynamic motion controller, which may necessitate limiting the drive torque in extreme cases to ensure sufficient torque and power for the steering. Furthermore, the wheel load and the speed of the electric motors must also be considered for the performance and stability of the corner module controller.

A further compromise must be found in the design of the gear ratios used. The gearbox connected to the actuators determines the maximum steering torque and the steering speed. Based on this, the bevel gear ratio can transform the torque and speed for the wheel. However, the package in particular limits the possibilities for the design of the bevel gearbox and its transmission ratio. Alternatively, the mechanical concept would have to be supplemented by an additional gearbox on the output of the bevel gearbox. This concept is therefore particularly suitable for vehicles whose vehicle speed is not excessively high in relation to the steering speed.

The prototype, which has a size and an output of up to 2 kW, was developed for an L7E vehicle. Due to the limitations of the available parts and budget, some compromises were made in the design. This is particularly evident in the selection of the bevel gearbox, the electric motors, and the rim. Another example is that only hall sensors and no high-resolution encoders were installed in the actuators. Consequently, control from a standstill and at low drive speeds and steering speeds is inaccurate, resulting in a loss of performance. In the absence of the aforementioned limitations, a more compact module would be feasible. The configuration depicted yields a maximum steering angle of ±55° for the prototype.

Two microcontroller development boards with real-time capability are employed to regulate and control the entire system. These boards receive all the necessary sensor data and control all the actuators. For the initial tests, the main voltage was reduced from 48 V to 36 V due to the limited specifications of the inverters. The control system was designed in Simulink.

The results demonstrate that this concept is both controllable and stable. The superimposition of steering and driving functions effectively. No cross interference is evident in the initial tests. The steering angle can be adjusted and maintained with an accuracy of 0.03°. This value increases slightly with higher loads. The step steer indicates that the existing control is a robust foundation upon which further enhancements can be constructed. The addition of encoders to the actuators is anticipated to enhance the control quality.

In light of these findings, further research is required to fully assess the potential of this concept. In addition to the design of a functioning prototype of the L7e class with four corner modules, the following research questions should be addressed:

- For which applications and vehicles is such a concept suitable?
- How much power is possible with high-voltage actuators when integrating the entire system into a 20" wheel rim?
- How can an electric actuated brake be integrated?

References

1. Feigel, H.-J.: Is Micro Mobility the Path Toward Corner Modules? In: Pfeffer, P. E., 14th International Munich Chassis Symposium 2023. Springer Vieweg, Berlin, Heidelberg (2023)
2. Struth, M., Faßbender, S., Sandmann, T., Schwarz, B., Eckstein, L.: Re-Defining Driving Experience - Competences & Concepts Behind the Research Vehicle SpeedE. In: 25th Aachen Colloquium Automobile and Engine Technologie, pp. 785-804. Aachen (2016)
3. Laurent, D., Sebe, M., Walser, D.: Assembly comprising a wheel and a suspension integrated into the wheel. U.S. Patent, 6,113,119, 2000/09/05
4. Zetterstrom, S.: Electromechanical steering, suspension, drive and brake modules. In: 2002 IEEE 56th Vehicular Technology Conference proceedings, pp. 1856-1863. IEEE, Vancouver (2002)
5. Gombert, B., Gutenberg, P.: Die elektronische Keilbremse. ATZ - Automobiltechnische Zeitschrift 108 (11), pp. 904-913 (2006)
6. Jonasson, M., Zetterström, S., Stensson Trigell, A.: Autonomous corner modules as an enabler for new vehicle chassis solutions. FISITA 2006 World Automotive Congress, FISITA Transaction (2006)
7. Yüksel, M., Ahmed, M., Girault, B., Birnschein, T., Kirchner, F.: A Framework for Design, Test, and Validation of Electric Car Modules. In: Fischer-Wolfarth, M. (eds.), 2014 Advanced microsystems for automotive applications, pp. 245-254. (2014)
8. Kraus, M., Harkort, C., Wuebbolt-Gorbatenko, B., Laumann, M.: Verschmelzung von Antrieb und Fahrwerk für einen People Mover. ATZ - Automobiltechnische Zeitschrift 120 (7-8), pp. 48-53 (2018)
9. Rajaie, M.-A., Khajepour, A., Pazooki, A., Soltani, A.: Development of A Novel Integrated Corner Module for Narrow Urban Vehicles. In: Proceedings of the Institution of Mechanical Engineers, Part D: Journal of Automobile Engineering, vol. 3, pp. 548-556. (2019)
10. Hyundai Mobis: Hyundai Mobis' e-Corner System Featuring Crab Walking and Zero Turn Becomes the First in the World to Drive on Public Roads. Press Release 2023/04/21. Seoul (2023)

11. REE Automotive, Website: REECorner | Modular Electric Vehicle Platform, https://ree.auto/technology/ , last accessed 2024/02/28
12. Martens, T., Pouansi Majiade, L.B., Li, M., Henkel, N., Eckstein, L., Wieglos, S., Schlupek, M.: UNICARagil Dynamics Module. In: 29th Aachen Colloquium Sustainable Mobility, pp. 863-884. Aachen (2020)
13. Hesse, L., Schwarz, B., Klein, M., Eckstein, L.: The Wheel-Individually Steerable Front Axle of the Research Vehicle SpeedE. In: 22nd Aachen Colloquium Automobile and Engine Technology, pp. 619-636. Aachen (2013)
14. Schwarz, B.: Auslegung einer energieeffizienten Vorderachse für ein radindividuelles Steer-by-Wire System. Doctoral thesis at Institute for Automotive Engineering – RWTH Aachen University. fka GmbH, Aachen (2020)
15. Eckstein, L.: Antriebsmodul mit integrierter Lenkfunktion für ein Kraftfahrzeug. German Patent Application DE 10 2016 007 445 A1, 2016/06/18
16. Böttcher, G., Böttcher, T., Eckstein, L., Latfullin, R., Reff, B., Carrillo Vasquez, C.: Potential of Road Surface Replicas for Use on Laboratory Test Rigs. In:12th International Munich Chassis Symposium 2021. Proceedings. Springer Vieweg, Berlin, Heidelberg (2021)

Standardization for a Steer-by-Wire vehicle

General disclaimer: The information provided below represents the current status of the activities and is subject to change based on further developments.

Alexander Ein Waldt, Ford-Werke GmbH
Technical Expert Steering Systems,
Spessartstraße 50735 Köln,
mail: aeinwald@ford.com

Oliver Nehls, Ford-Werke GmbH
Controls Engineer Steering Systems
Spessartstraße 50735 Köln
mail: onehls@ford.com

Goetz-Philipp Wegner, Ford-Werke GmbH
Functional Safety Engineer Steering Systems
Spessartstraße 50735 Köln
mail: gwegner2@ford.com

P. E. Pfeffer (Ed.), *15th International Munich Chassis Symposium 2024*,
Proceedings, https://doi.org/10.1007/978-3-662-71064-7_11

Abstract

The future technology Steer-by-Wire enables a variety of additional features in vehicle use. The spectrum ranges from extended driving dynamics functions, autonomous driving to new interior concepts. However, there has been uncertainty in the industry as to what requirements are necessary for a safe implementation of this new technology.

Both sets of current standards, the DIN 70065 and ECE-R79 cover explicitly and implicitly performance requirements for normal operation and requirements in case of a fault. These requirements focus on the SbW system, but also supporting systems in the vehicle like power supply or brake systems are affected. Additionally, the ISO26262 provides a generic approach to reduce risks to a reasonable level.

This technical paper investigates how the present standards cover for the demand to develop safe Steer-by-Wire and supporting system designs, identifies potential gaps, and proposes the potential need of advancing those standards further. Aligning the industry to a common requirement landscape for SbW will ensure that the required steering performance and potential vehicle limitations are uniformly managed and implemented so that vehicle behavior is consistent to the customer.

Keywords:
Steer-by-Wire, Standards, Safe State, Availability, Safety, Degradation, Transition, Minimal Risk Maneuver

1 Introduction

The market penetration of by-wire systems (Brake-by-Wire/Steer-by-Wire) differs fundamentally from current active chassis systems such as Electrical Power Steering (EPS) or Continuous Control Damping (CCD) in that the by-wire safe states have more stringent requirements for fail-operational designs.

Therefore, it is crucial to gain a more precise insight into three fundamental aspects:

- Availability criteria must ensure minimal functionality even after initial faults occur.
- The resulting safety concepts depend on vehicle-level interventions to guarantee the vehicle reaches a safe-state, including actions like deceleration or immobilization in accordance with the 'transition to safe state' protocols.
- Level 3 and higher autonomous vehicles follow a similar strategy by implementing Minimal Risk Maneuvers (MRM) in response to faults.

Unlike Level 3+ autonomous vehicles, where the driver is no longer in the control loop, in this case, the driver remains in control. Therefore, a Minimal Risk Maneuver (MRM) cannot be independently defined to reach a safe state. Since the driver is still in control, the definition and development of MRMs must consider the driver's perspective to ensure that they can safely operate the vehicle until a safe state is achieved.

The current standards have not provided sufficient guidance how to design for the critical aspects of availability, vehicle level safe state and the transition to the safe states. As a result, new standards such as DIN 70065 have been and ISO 19725 are being created, and existing standards like UNECE-R79 are being proposed to be enhanced to ensure consistent guiding principles across the industry.

The paper will start by examining the existing standards (UNECE-R79 and ISO 26262) to outline what is currently being considered in relation to these three areas. It will then demonstrate how the recently established DIN standard for Steer-by-Wire addresses the criteria for safe states and their transition, along with any underlying requirements for other vehicle systems.

Furthermore, a brief overview of the suggestions for aligning the existing UNECE-R79 standard will be provided. Additionally, there will be a discussion on the future objectives of the presently developed Steer-by-Wire ISO 19725.

The document concludes with a look ahead at particular aspects that should be considered for incorporation into future standardization efforts to establish a common stance across the industry.

2 Present standards

Presently, two standards have a direct impact on the architecture and design of a Steer-by-Wire System (SbW).

The ISO 26262 [1] offers a framework with methodologies and tools to attain a safe system design and to reduce risks to a reasonable level. If an SbW-System is developed in accordance with ISO 26262, a logical implication is that the Steering System must be a Fail-Operational System. This entails that the vehicle's steerability necessitates at least one level of redundancy. ISO 26262 provides flexibility in achieving the necessary redundancies for steerability. Consequently, the market could offer a wide range of solutions with high variability of how customers experience the vehicle's response following a fault.

The ISO 262262 offers limited guidance in the following areas:

1. The system or vehicle's post-fault state should align with the remaining capabilities and the current driving situation. For instance, if the steering system retains high authority and redundancy after a fault, it can continue operating. In other words, as a generic standard, ISO 26262 does not specify the necessary design space or framework beyond safety for a SbW-System design.

2. The transition requirements for the safe state within ISO 26262 are unspecific, primarily emphasizing achieving safety within the Emergency Operation Tolerance Time Interval (EOTTI). The design approach for transitioning the vehicle to a safe state can vary among different OEM's. This variability can result in inconsistent behaviors across OEM's and for their customers. For instance, the responses to sudden loss of assist in EPS have varied among OEM's, raising concerns in the field.

3. Even if the transition to the defined safe state complies with standards, both the transition process and the safe state itself may be perceived as unsafe by the customer. For instance, abruptly stopping the vehicle even when some level of steerability is still possible could result in an unfavorable stopping position.

4. ISO 26262 does not specify requirements for the availability of the steering system as long as it is safe. The standard does not address how to enhance availability or related considerations to prevent customer dissatisfaction. This lack of coverage could result in reduced customer acceptance of the technology.

The UNECE-R79 [2] regulation for steering equipment sets out the vehicle-level requirements that a steering system must meet for vehicle homologation. The UNECE-R79 has established requirements for a Steer-by-Wire system, also known as "full power steering." These requirements were formulated prior to the widespread accumulation of serial development experiences. Furthermore, certain technological aspects such as centralized control modules, integrated vehicle dynamics controls, highly redundant systems, and Battery Electric Vehicles have advanced rapidly in recent years and were not considered during the development of that standard.

Overall, the following elements in UNECE-R79 may lead to discrepancies with current design approaches:

1. UNECE-R79 solely addresses requirements for initial fault responses, lacking provisions for multiple faults and highly redundant systems.

2. A red warning for (first) fault is defined. That limits the opportunity to use yellow warning light if sufficient redundancy is still available. That can unnecessarily limit the availability of the vehicle for the customer.

3. Based on the above two points, the benefits of highly redundant system architectures can only be partially realized.

4. The requirements for a redundant power supply are implicitly outlined.

5. A general requirement prohibits driving at higher speeds indefinitely after a fault, but the specific time limit is not defined. UNECE-R79 implicitly assumes that driving at a maximum of 10 km/h is consistently safe. In scenarios involving consecutive failures (such as shared faults between steering and Brake-by-Wire systems), halting the vehicle may be the preferred vehicle state.

6. A comprehensive safety concept at the vehicle level for an SbW system, including other systems like the powertrain and brake system, regarding safe states, degradations, and transitions to safe states has not been considered. The existing regulation thus restricts the ability to implement state-of-the-art architectures and corresponding vehicle-level safety concepts. This limitation can impact availability and may

necessitate additional design efforts, such as requiring a large energy reservoir instead of a more intelligent vehicle-level degradation approach.

7. The performance requirements provided are rooted in traditional steering systems and may be revised to align with SbW-specific use cases.

With the growing implementation of mechatronic systems and by-wire technologies, along with the increased interconnectivity of these systems, regulations and safety cases at the component level progressively reach their limits.

Furthermore, the capacity to implement advanced solutions at vehicle level, facilitated by multiple actuator systems, is not addressed within a component-only framework. To address these limitations, current standards must be adapted to enable a comprehensive multi-component approach.

Moreover, new standards specifically tailored for SbW systems are currently in development.

3 Revised standards – UNECE-R79

Considering the constraints mentioned earlier, the industry has suggested revising the regulatory provisions. In this paper, two key amendments will be introduced.

A more granular warning strategy enables the utilization of a yellow warning light under specific conditions following an initial fault. This approach empowers the user to assess the level of redundancies in their SbW system and at vehicle level. If, post-fault, there is at least one redundancy for safe vehicle steering (consisting of two or more steering channels), a yellow warning light suffices, as a subsequent failure would not lead to loss of steerability. **Fig. 1** illustrates a comparison between the existing standard and the proposed modification [3].

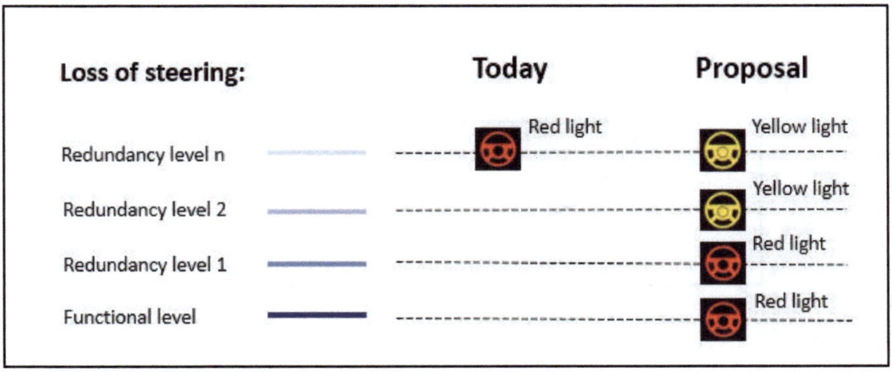

Fig. 1. Proposal for enhanced warning requirements based on the distance to loss of steerability

Since the current regulation does not consider a vehicle-level degradation to reach a safe state (such as automated speed reduction), the standard specifies a fixed energy reservoir in the event of an energy source failure. This energy reserve should provide adequate time for the driver to safely stop the vehicle. This method has some implications:

- It depends on the driver's correct and timely reaction to halt the vehicle before the energy reservoir is depleted.
- Any other power consumer might drain the energy reserve more rapidly like climatization or operation under severe environmental situations (e.g. very low temperatures).
- Having an independent reservoir may lead to a substantial increase in capacity or size compared to conventional steering systems.

The alternative method proposed aims to address failures in the energy source. In the event of an energy source fault, the system can automatically initiate deceleration to bring the vehicle to a stop. The vehicle must come to a standstill no sooner than 60 seconds. The driver retains the ability to independently stop the car at any time.

A revised definition for the minimum energy reservoir is necessary. A sinusoidal steering input will be applied during the complete deceleration phase to ensure the driver can steer safely. If braking and steering still have separate energy sources post-fault, decelerating to 10 km/h is sufficient as it allows the vehicle to steer or brake in the event of another fault. If the same energy source for steering and braking is compromised, the vehicle must come to a standstill, as depicted in **Fig. 2** [3].

The chart demonstrates various methods to safely bring the vehicle to a stop depending on the traffic conditions.

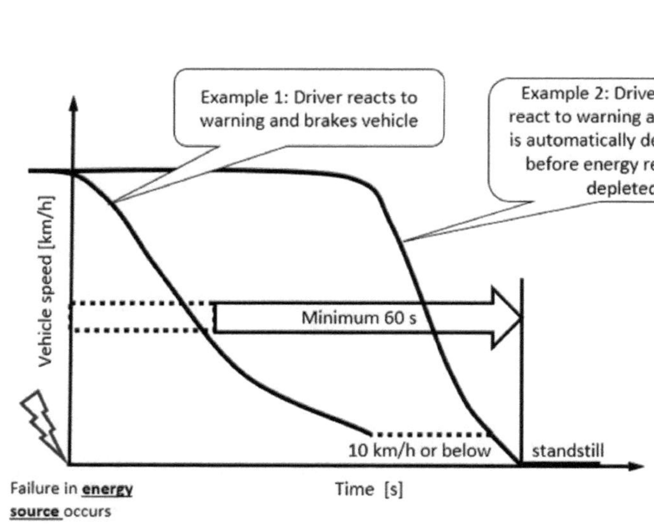

Fig. 2. Proposal for an alternative to reach safe state of vehicle (for shared power source for braking and steering)

Given the steering-related nature of this requirement, the brake system must establish its own safety case.

By adjusting the start of deceleration based on the energy reservoir's actual level, this approach maximizes the driver's potential drive time to reach a suitable stopping point.

4 New standard – SbW DIN70065

Due to growing industry interest in SbW technology, there has been a need to resolve the uncertainty regarding the implementation of this technology in vehicles, particularly in terms of safety, availability, and degradation in the event of a fault. Hence, a collaborative project involving OEM's and suppliers under the VDA's and DIN's auspices was initiated to create a new industry-wide standard.

The emphasis is on addressing key topics of general interest, including item definition, key safety goals, error categories, controllability following an initial fault, operational behavior post-first fault, and the related degradation concept including a set of verification tests.

A degradation concept is essential for Steer-by-Wire applications to ensure a seamless transition of the vehicle to a safe state. In this safe state, the Steer-by-Wire vehicle operates at low speed in "Creep Home" mode or comes to a complete stop in "Vehicle Standstill" mode.

To enable continuous driving following an initial fault, the DIN standard's degradation concept outlines requirements for various degradation levels. Specifications for steering capabilities and the remaining safety integrity are detailed for these degradation levels, along with guidelines for transitioning between these states.

A maneuver catalog with corresponding acceptance criteria has been developed to guarantee adequate steering capabilities at the vehicle level in fault scenarios.

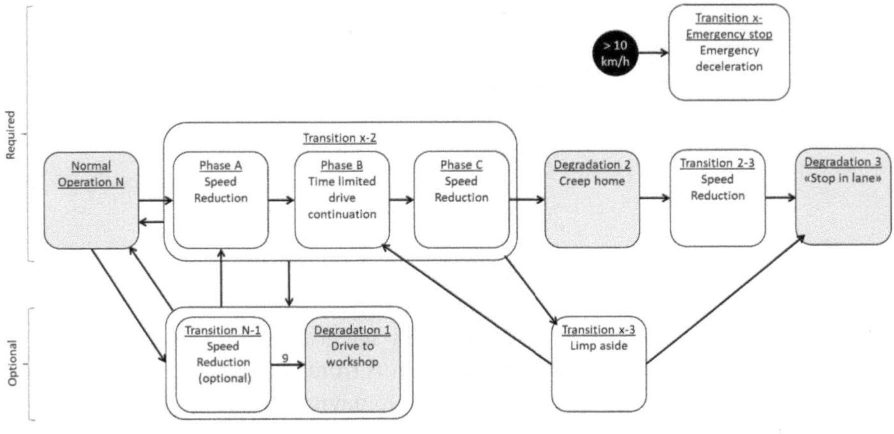

Fig. 3. Principle overview of the degradation concept

As depicted in **Fig. 3** [4], a fundamental set of states and degradations has been established to ensure the safe degradation of the vehicle following one or more failures, leading to the designated safe states (middle row boxes). The lower path boxes indicate two optional states for systems with higher redundancies. Furthermore, the standard specifies, beyond the ISO 26262 standard, the appropriate vehicle response in the event of a safety goal violation (loss of steerability).

The degradation concept establishes requirements, among others, for supporting vehicle systems, the safety integrity of the steering system, the warning concept, and vehicle lateral control.

These requirements are categorized as follows:

- System integrity and speed limitations
- Safety integrity
- Restrictions on duration of continued operation
- Recovery and its conditions
- Warning concept
- Minimum requirements for lateral control and controllability of the vehicle

Supplementary guidelines regarding transition times for maintaining driving with higher speed after a fault are outlined. These guidelines complement the proposed amendments to UNECE R-79 regarding the minimum energy reserve following an energy source failure.

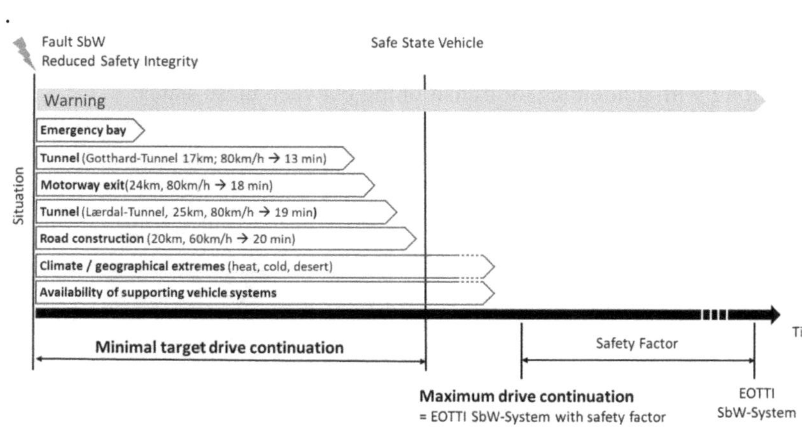

Fig. 4. Drive continuation time with reduced safety integrity

The guidelines for driving with reduced safety integrity, as shown in **Fig. 4** [4] (e.g., after the loss of redundancy), consider time limits from two perspectives. On one hand, they outline the durations needed to continue to navigate the vehicle in common scenarios where the driver may prefer not to bring the vehicle to a halt. For instance, the availability of convenient stopping locations within a tunnel is restricted.

On the other hand, instructions are given on the maximum duration for driving continuation without elevating the risk of a subsequent failure to an unacceptable level. The guidance commences with an EOTTI time as specified in ISO 26262. To accommodate uncertainties in the EOTTI calculation and undetected dependent faults (such as pre-damaged components due to overheating), a safety factor is introduced.

Establishing a unified vehicle degradation concept across the industry will guarantee that fault handling and potential vehicle performance restrictions are consistently addressed and executed, ensuring uniform vehicle behavior for consumers.

For further information, refer to the technical paper and presentation from the chassis.tech plus conference held in 2023 (Basic safety guidelines for Steer-by-Wire for a new DIN standard).

5 Evolution of standards – ISO 19725

To expand the spread of application, the development of an ISO standard began in the autumn of 2023. The ISO 19725 draft, with a general focus on safety and availability, is modeled after the DIN 70065. This alignment with the DIN serves as the foundation for the initial draft of the ISO, facilitating the seamless transfer of valuable expertise accumulated over the recent years.

Ultimately, a comprehensive technical representation encompassing key automotive industry countries such as the USA, China, Japan, Korea, and various European nations guarantees that the specific requirements of each market are duly accounted for.

6 Discussion

While the DIN standard focuses on safety and availability, and the UNECE-R79 is undergoing updates to resolve conflicting requirements, there are still some areas that require further exploration for future development.

One of the key unresolved issues for the future is to comprehend the implications of pattern of defect. The automated deceleration of the vehicle in response to steering faults introduces a novel behavior for customers. While this reaction is deemed appropriate from a safety standpoint, customers may perceive it as potentially unsafe. Hence, it is essential to determine the frequency at which this vehicle response can be tolerated across the OEM's fleets.

Additionally, gaining more experience is necessary to identify the most effective means of informing customers and determining the optimal deceleration levels. Similarly, the warning strategy for continued driving following faults (refer to Fig. 3, TRx-2) necessitates extended field feedback for potential adjustments and adaptations.

The advent of Automated Driving (L3+) necessitates a thorough analysis and comprehension of the differences and similarities between the L3+ system and the SbW-System's responses to faults (degradation concepts).

In the SbW-System, where the driver is always actively involved, the goal is to prolong the driving time after faults to allow the driver broad opportunity to locate a suitable parking spot. In contrast, the L3+ System provides only minimal handover time to the driver following a fault, transitioning swiftly to a Minimal Risk Maneuver (MRM) such as stopping in the lane or performing a limp-aside maneuver. This restricted availability is attributed to the inherent limitations in sensor redundancy.

Ultimately, there is a need for clarification within the industry, particularly between the Chassis and Power Supply Systems, to determine if supplementary requirements are needed during the vehicle's transition to a safe state (standstill).

This may entail an analysis to ascertain whether the current requirements adequately address these scenarios. The current requirements already stipulate a minimum energy reservoir (refer to UNECE-R79) and the necessary peak power draw (refer to UNECE-R79 and DIN 70065). The remaining question is whether the industry should establish a consensus on an additional standardized maneuver for power supply dimensioning.

7 Summary

The DIN 70065 aims to substantiate the application of ISO 26262 for a SbW System. Furthermore, it includes supplementary requirements to improve system availability and define the degradation of a SbW-System, aiming to enhance customer acceptance post faults.

The new standard offers a robust framework for verifying the SbW design, encompassing first fault mitigation, minimal performance across various system states, and the transition and degradation concept. The ongoing development of ISO 19725 aims at creating an international standard by incorporating the global expertise of participating entities. These initiatives have prompted a re-evaluation of the UNECE-R79 and a proposed adaptation to align it with current and future standards.

A significant result of these standardization efforts is a more consistent development process enabled by a globally aligned requirement landscape. These standards can also serve as the foundation for a universal basis for homologation across various regions.

Aligning the industry on those standards, coupled with clear availability and performance requirements post faults, mitigates the risk of field-related concerns.

The factors mentioned above can significantly enhance the acceptance of SbW-technology, facilitating the rapid adaption and implementation of this innovative steering system and its associated customer benefits in the market.

References

1. International Organization for Standardization: ISO 26262:2018 Road Vehicles – Functional Safety (2018)

2. United Nations: Addendum 78: UN Regulation No. 79: Uniform provisions concerning the approval of vehicles with regard to steering equipment (Rev.1/Add.78/Rev.5/Amend.4, 2024)

3. Proposal for Supplements to the 03 and 04 series of amendments to UN Regulation No. 79, documents GRVA-18-17 and GRVA-18-19. In: https://unece.org/info/Transport/Vehicle-Regulations/events/384684 (Accessed March 18th, 2024)

4. Dr. Schölzel, M., Kleiner, C., Ein Waldt, A.: Basic safety guidelines for Steer-by-Wire for a new DIN standard. In: Pfeffer, P. (eds) 14th International Munich Chassis Symposium 2023. IMCS 2023. Proceedings. Springer Vieweg, Berlin, Heidelberg. Publishing date 06.2024. DOI will follow.

Driving simulator study on the acceptance of take-over concepts of a Steer-by-Wire functionality

Hendryk Lausch[1], Tim Ahrenhold[2] and Roman Henze[1]

[1] Technical University Braunschweig, Institute of Automotive Engineering (IAE), Hans-Sommer-Straße 4, 38106 Braunschweig, Germany
[2] IAV GmbH, Rockwellstraße 3, 38518 Gifhorn, Germany

Abstract. Innovative Steer-by-Wire steering systems (SbW) offer an enormously increased functional potential compared to conventional EPS steering systems, especially on the way from partially automated driving to highly and fully automated or even autonomous driving (SAE Level 3 and 4 or 5). ADAS play an increasingly important role in today's world of more and more complex and unmanageable traffic scenarios. In order to fulfill the requirements of the Vienna Convention it is mandatory that ADAS for assisted and automated driving are designed to be overrideble by the driver. Therefore, this paper investigates different approaches for a cooperative automation to regain vehicle control by the driver during an ADAS intervention. The basis is a SbW steering model created in advance, which was validated by measurements at the institute's own Hardware-in-the-Loop steering test bench (HiL). The considered ADAS describes an innovative SbW function for automated evasion, which initiates the steering maneuver without the driver's intervention. This driving function is extended by a take-over function, which is considered in different application variants by parameter variation. Both threshold values and the transition behavior between system and driver influence are varied. The focus of this paper is a subsequent subject study on the institute's own driving simulator (Dynamic Vehicle Road Simulator, DVRS), which examines and evaluates the various take-over applications. Subjective criteria for the take-over behavior as well as objective characteristic parameters of vehicle dynamics quantities are included in the evaluation of the application variants.

Keywords: Steer-by-Wire steering system, Autonomous evasion function, Advanced driver assistance systems

1 Introduction

In today's world where the traffic scenarios become more and more complex and hard to handle, modern advanced driver assistance systems (ADAS) are already able to reduce the number of accidents with a potential of 70% [1]. However, this potential could be increased by the use of highly innovative Steer-by-Wire steering systems (SbW) as these systems represents a completely new steering concept with mechanical decoupling of the steered front wheels and the steering wheel respectively the drivers

P. E. Pfeffer (Ed.), *15th International Munich Chassis Symposium 2024*,
Proceedings, https://doi.org/10.1007/978-3-662-71064-7_12

action. This enables new approaches or extensions to existing ADAS especially regarding those with higher lateral dynamics, as there is no longer a danger to the driver from a fast and suddenly turning steering wheel.

In this paper and as an example of the potential mentioned above, the evasion assistant is considered. In today's vehicles with conventional EPS steering systems, the evasion assistant is mostly implemented in a way, that as soon as an obstacle appears in front of the vehicle and a critical situation is detected by the sensors (radar, lidar, camera), the driver gets a visual warning and then needs to initiate the evasive maneuver by himself. The EPS system then provides an additional steering torque via the rack actuator to assist the drivers steering movement and reduce the required steering torque. If the driver wouldn't react to the situation the evasion trajectory would not be driven (only an autonomous braking assistant could reduce the collision speed). [2]

With the use of SbW this ADAS could be extended to an autonomous evasion function which initiates the evasive maneuver before the driver is able to react to the situation. This means the driver gets decoupled for a short time and the SbW function acts on its own to follow the ideal evasion trajectory, but the driver is then able to regain the lateral vehicle control as soon as he reacts to the situation.

For this a take-over strategy is needed to ensure the driver regains the vehicle control in a safe and comfortable way, ideally without even noticing that the evasive maneuver was actually initiated by the SbW system and not by the driver. Therefore, this paper investigates a possible way to hand over the vehicle control from the SbW system back to the driver in case of the autonomous evasion function, which was modeled and implemented beforehand.

The general toolchain that was used is shown in **Fig. 1**. Initially a real SbW system was measured on the institute's own Hardware-in-the-Loop (HiL) steering test bench to determine the transmission behavior in form of system parameters in order to recreate its characteristic by means of a steering model. In combination with a dual-track model it was applied on the driving simulator to investigate the autonomous evasion function respectively the various take-over applications in the study. The goal with the performed study is to get feedback on the general acceptance of the take-over strategy and how the varied parameters influence the rating compared to a reference variant, where the participants did the evasive maneuver without any assistance. Among others the subjects were asked to evaluate the haptic steering feeling, the subjective feeling of safety and controllability of the vehicle.

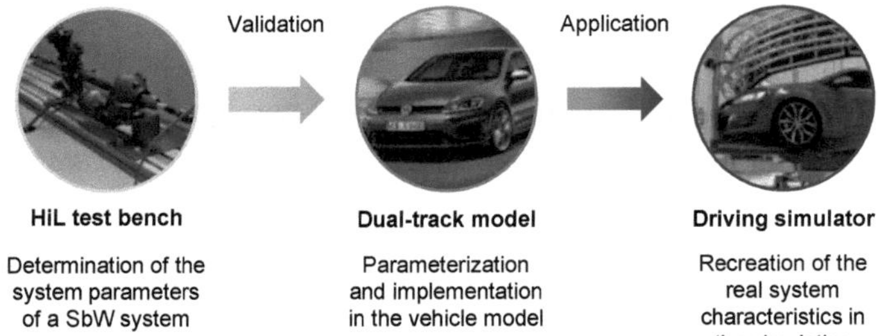

Fig. 1. Toolchain of the driving simulator study

2 Simulative braking and steering under low road friction conditions

2.1 Conflict of braking/steering

This subsection describes the simulative investigation of the conflict of objectives between braking and evasion as options for preventing a collision of the ego vehicle with an obstacle to outline the potential of an autonomous SbW evasion function. Furthermore, the driving maneuver on which the study was based on is explained.

When it comes to avoiding a collision with an obstacle, there are basically two options: braking or evading (or both combined). However, depending on the road conditions, the relative speed and the distance of the ego vehicle to the obstacle, not both options can always prevent a collision. In some cases, a crash can only be avoided by either evading or braking. This leads to a conflict of objectives, which is investigated within a pre-simulation presented in the following.

The simulation considered braking and evasion for different relative speeds between the ego vehicle and the obstacle as well as the road friction coefficient. As results the minimum distance to avoid a collision was identified for both braking and steering. If this necessary distance is plotted over the relative velocity the curves for the "Last Point to Brake" (LPTB) and the "Last Point to Steer" (LPTS) are obtained. **Fig.2** shows these two curves exemplary for dry road conditions (μ_R=1).

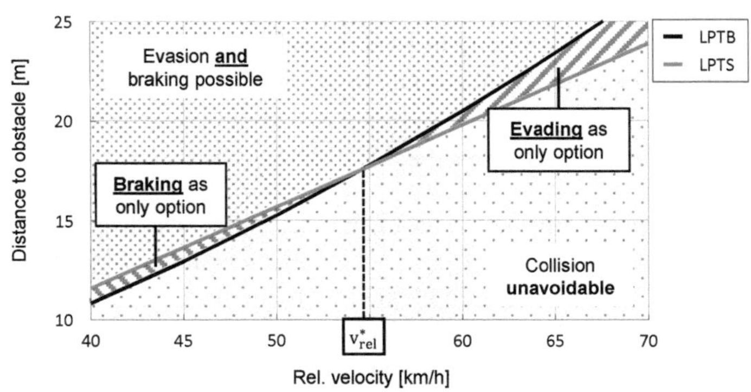

Fig.2. LPTB (black curve) and LPTS (grey curve) for a road frictions coefficient of 1.

The black curve indicates the LPTB, the grey curve presents the LPTS. As shown in the illustration, depending on the driving situation only one of the options evasion/braking would possibly avoid a collision (marked with the hatched areas). These two areas are characterized by the intersection of the two curves at a relative velocity of v_{rel}^*. For $v_{rel} > v_{rel}^*$ the LPTS is "behind" the LPTB ("closer" to the obstacle), which means with decreasing distance to the object at some point braking would no longer avoid the collision and evasion becomes the only option (which is indicated by the area between the two curves). This area also marks the relevant area in which an autonomous SbW evasion function could intervene and unfold its potential of increasing driving safety. For $v_{rel} < v_{rel}^*$ it is the other way around (LPTB is "closer" to the obstacle than the

LPTS) and with decreasing distance to the obstacle braking becomes the only option to avoid a collision and evasion is no longer effective.

The LPTB corresponds to the required braking distance s_B and can be described with the following formula F1.1 [3]:

$$s_B = \frac{v_{rel}^2}{2 * a_{x,max}} + v_{rel} * T_{BLT} \qquad \textbf{F1.1}$$

Assuming a still standing obstacle (vehicle), the relative speed v_{rel} equals the speed of the ego vehicle, on which the braking distance is squarely dependent. T_{BLT} is the brake loss time and considers the effective time loss from pressing the brake pedal until the maximum braking force is applied. This time results of the inertias and clearances in joints and bearings of the braking system components and the built-up of the braking forces [4]. The maximal achievable longitudinal deceleration $a_{x,max}$ is the product of the road friction coefficient μ_R and g according to F1.2 and is therefore determined by the boundary conditions:

$$a_{x,max} = \mu_R * g \qquad \textbf{F1.2}$$

For the LPTS, a 7th degree Polynomial was used to approach the evasion trajectory with a lateral offset d_{lat} of 2.5 m in combination with a defined lateral acceleration a_y (double derivation of the given offset) for different vehicle speeds. The defined lateral offset d_{lat} results from an assumed vehicle width of 2 m (VW Golf 7) and a lateral safety distance of 0.5 m. Analog to F1.2 the achievable lateral acceleration $a_{y,max}$ in general is defined as in F1.3. According to [4] the factor k is between 1 and 0.8.

$$a_{y,max} = \mu_R * g * k = a_{x,max} * k \qquad \textbf{F1.3}$$

The LPTS depends on the length of the trajectory, which in turn depends on the achievable lateral acceleration $a_{y,max}$, which again is defined by the road friction co-efficient (see F1.3)). This means both the LPTB and the LPTS are highly dependent of the road conditions. The conflict of objectives for the representative road friction coef-ficients of dry (1.0), wet (0.6) and snowy (0.3) road conditions are shown in **Fig.3** [5]. The **Table 1** furthermore lists the selected parameters for the pre-simulation of the LPTS and the LPTB.

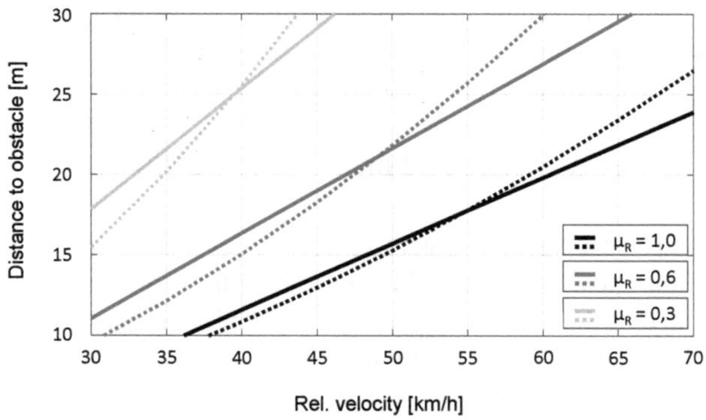

Fig.3. LPTB and LPTS for dry, wet and icy road conditions

Table 1. Selected parameters for simulative LPTB and LPTS

Parameter (LPTB)	**Value**
Brake loss time T_{BLT}	300 ms [6] [7]
Parameter (LPTS)	**Value**
Factor k	0.9 [4]
Lateral offset d_{lat}	2.5m
Parameter (Other)	**Value**
Road friction coefficient μ_R	[1.0, 0.6, 0.3]
Ego vehicle speed	[10:5:120 km/h]
Obstacle vehicle speed	0 km/h

As can be seen, with lower road friction coefficient the intersection point of the LPTB and LPTS moves to lower relative velocities, which means the potential of an SbW evasion function increases with worse road conditions (the area where evasion is the only option becomes larger and is not limited towards higher velocities).

Table 2 contains the final boundary conditions for the study scenario, which are selected based on the results of the pre-simulation. To reduce the dynamics of the driving simulator in the later study, the ego vehicle speed was set to v_{veh}=50 km/h. To ensure the constant velocity in the study, the participants used the cruise control function in the simulator. With the selected vehicle speed and regarding **Fig.3** the road friction coefficient was selected to μ_R=0.6 to represent a driving situation in which evasion would be the only option to avoid a collision, since this is the area with the highest SbW potential.

Table 2. Selected boundary conditions for the study scenario

Study	Value
Driving speed of ego vehicle v_{veh}	50 km/h (*urban*)
Road friction coefficient μ_R	0.6 (*wet*)

Since a road friction coefficient of $\mu_R = 0.6$ has been determined in the previous steps for one of the boundary conditions for the study scenario, the maximum lateral acceleration according to F1.4 results to:

$$a_{y,max} = \mu_R * g * k = 0.6 * 9.81\ m/s^2 * 0.9 = 5.297\ m/s^2 \qquad \textbf{F1.4}$$

The maximum drivable trajectory with this lateral acceleration and the defined lateral offset of 2.5 m is shown in **Fig.4** with a total length of $s_E = 25.95$m. This trajectory was therefore defined as the evasion trajectory for the SbW evasion function which was used in this paper and later extended with the take-over functionality.

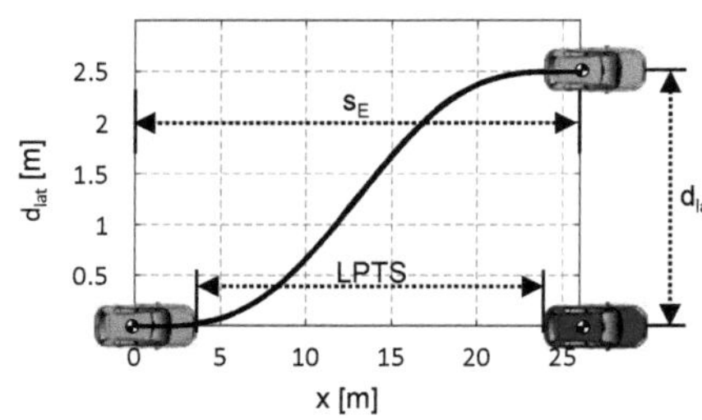

Fig.4. Defined trajectory for the evasive maneuver for the simulator study

2.2 Steer-by-Wire evasion function design

In this subsection the modeled SbW function is explained, which has been extended with the take-over concepts that are illustrated in 2.3.

For the modelling of the autonomous SbW evasion function an institute's own dual-track model has been used as a basis. This complete vehicle model includes a sub-model of a Steer-by-Wire steering system that has been validated at the Hardware-in-the-Loop (HiL) steering test bench with a series of measurements at the institute. The parameters of the SbW components (such as rack friction level, inertias of rack and pinion, steering ratio, etc.) were determined in such a way to simulate a real steering transmission behavior.

The autonomous evasion function is based on the evasion trajectory that was defined in 2.1 (**Fig.4**) as the target trajectory. For the realization a closed-loop control was combined with a feedforward (**Fig.5**). The yaw rate $\dot{\psi}$ and the yaw angle ψ were used as the two driving dynamics parameters to control the evasion trajectory. Controlling the yaw rate ensures a stable diving condition while controlling the yaw angle makes sure that the vehicle completes the trajectory in the longitudinal direction to the road. The actual values of $\dot{\psi}$ and ψ were provided by the dual-track model while the target parameters $\dot{\psi}_{target}$ and ψ_{target} are resulting of the target trajectory that was determined with the 7th degree Polynomial in the pre-simulation of the LPTS with the parameters listed in Tab.2 and shown in **Fig.4**. The yaw rate $\dot{\psi}$ and yaw angle ψ were both controlled by a simple PID-Controller.

For the feedforward the formula context of the self-steering gradient (SG) is used to determine a target rack position x_{rack,δ_S} that is needed for the desired target yaw rate $\dot{\psi}_{target}$. For this, a target steering wheel angle δ_S is determined which is then converted into a rack position via the steering gear ratio i_{Rack} (determined within the HiL measurements). The three parts of the yaw angle control, the yaw rate control and the feedforward provide the control variable of the target rack position $x_{rack,target}$, which is fed back as input into the steering model. Within the SbW model, the target rack position is controlled using a torque control that determines the target torque of the rack actuator.

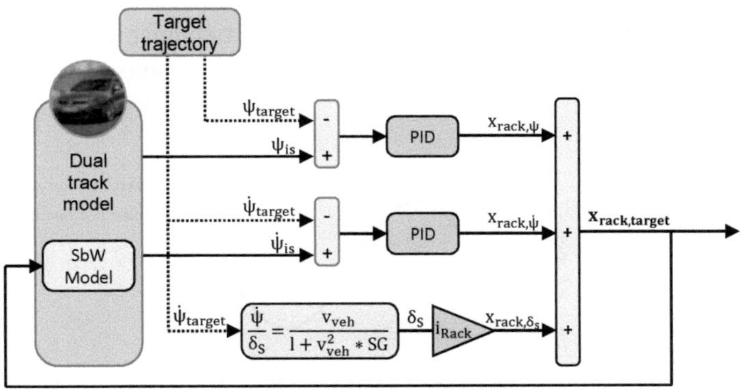

Fig.5. Modelling of the evasion trajectory

2.3 Take-Over modelling

In this subsection the concept of the modeled take-over strategy is explained.

This paper makes the assumption of ideal sensors with no further consideration of any latency to detect the critical situation or any fault detection. This means the evasion trajectory is initiated as soon as the LPTS related to the obstacle is reached and therefore earlier than the human driver would react to the situation. This way was chosen to investigate the maximum potential that could be achieved with such a function. This also means the driver is fully decoupled from the lateral vehicle control at the beginning of the evasion trajectory but is able to regain the control afterwards. For this the take-over concept considers two steps. First, it has to be decided if and when the driver receives back the control of the vehicle. And second, the transition characteristic between the system-side and the driver-side part of the vehicle control has to be modeled.

The idea of the take-over concept used for the study presented in this paper is to consider the deviation of a predicted driving condition based on the driver's steering wheel input and the ideal (actual) driving condition based on the driven trajectory of the SbW evasion function to decide, if and when the lateral vehicle control is transferred back to the driver. To ensure that the driver only regains control of the vehicle if he or she reacts appropriately to the driving situation (e.g. no overreacting) and the controllability of the vehicle is guaranteed, three criteria were defined that all must be met at the same time for a certain period of time. Each criterion uses a specific vehicle dynamic parameter to define boundaries that limit its deviation of the predicted driving condition (driver) to the ideal vehicle behavior (SbW function). These three criteria are the deviation of the lateral offset Δd_{lat}, the deviation of the yaw rate $\Delta\dot{\psi}$ and the deviation of the float angle $\Delta\beta$. For each criterion an upper and lower limit was defined. While Δd_{lat} ensures that the vehicle would follow the target trajectory and would be able to reach the required lateral offset within the required longitudinal distance to safely drive around the obstacle if the driver would receive back the control of the vehicle, $\Delta\dot{\psi}$ and $\Delta\beta$ are parameters to describe the vehicle stability to ensure a take-over would not result in an unstable and uncontrollable driving behavior. The limits for $\Delta\dot{\psi}$ and $\Delta\beta$ were the same for all investigated application variants. The tolerance range for the yaw rate is shown in the **Fig.6**. It is defined as a tolerance band which allows a deviation of ±0.2 rad/s compared to (ideal) yaw rate that results of the SbW evasion function.

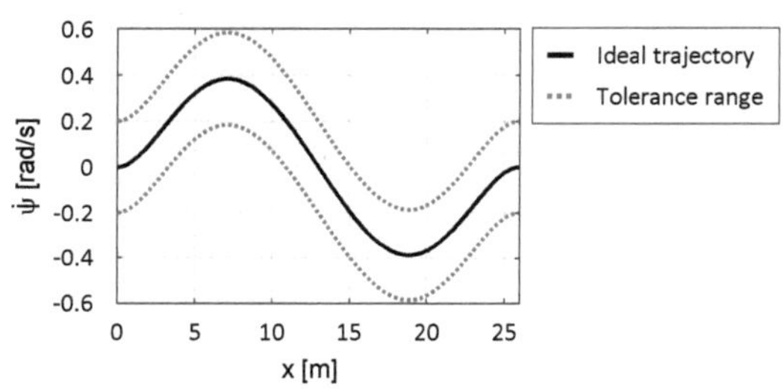

Fig.6. Yaw rate of the ideal trajectory and the upper and lower limit for the deviation of the yaw rate

For $\Delta\beta$ the limits were set at constant values so the condition -0.0873 rad $< \beta <$ 0.0873 rad needed to be fulfilled since for a floating angle in this range a stable driving behavior can be assumed [6]. The upper and lower limits for the deviation of the lateral offset Δd_{lat} were varied through the application variants in form of two tolerance bands like shown in **Fig.7**.

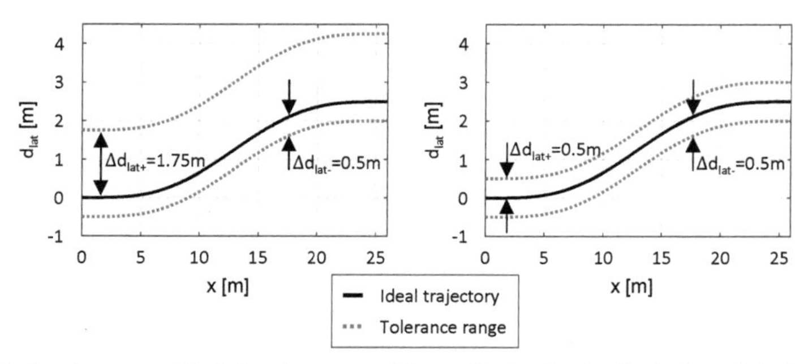

Fig.7. Ideal trajectory with defined upper and lower limits for the deviation of the lateral offset (left: bigger tolerance band, right: smaller tolerance band)

For the study two different types of tolerance bands for the deviation of the lateral offset were defined assuming a two-lane road with a lane width of 3.5 m and a vehicle width of 2 m. One is designed as a "bigger" tolerance range with an upper limit of +1.75 m and a lower limit of -0.5 m deviation to the evasion trajectory (meaning a permissible total lateral offset of 2 m to the right and 4.25 m to the left). The upper limit results of the lane boundary as the vehicle would leave the road if it would exceed a lateral offset of 4.25 m. The lower limit of the tolerance band with a total lateral offset of 2 m equals the required offset to drive around the obstacle vehicle without a crash. For the "smaller" tolerance range the mentioned lower limit of -0.5 m deviation is also applied to the upper limit to receive a symmetric tolerance band of ±0.5 m. This results in a permissible total lateral offset of 3 m to the left and 2 m to the right.

All three criteria mentioned ($\Delta\dot{\psi}$, $\Delta\beta$ and Δd_{lat}) need to be fulfilled for a certain time to safely transfer the vehicle control back to the driver. Therefore, a dwell time t_{dwell} has also been defined and varied through the application variants. The dwell time defines how long these criteria needs to be met simultaneously to let the driver regain the control of the vehicle. A longer dwell time means longer decoupling of the driver (as the SbW function remains active in this period of time, which could generate a negative feeling of an uncontrollable or less controllable vehicle) but at the same time ensures that the predicted driving condition based on the driver's behavior maintains within the given boundaries for a longer period of time. A shorter dwell time let the driver regain the vehicle control back earlier (which again could result in a positive subjective feeling that the driver never relinquished control or he didn't even notice the intervention of the evasion function) but on the other hand could still result in a less stable driving condition after the take-over since the criteria needs to be fulfilled for a shorter period of time. Regarding the study two different types of tolerance bands (small and large) for the deviation of the lateral offset as well as two different dwell times (short and long) were investigated in this study (**Table 3**) which were then combined into four different application variants V1 to V4 like described in 4.1 (**Table 4**).

Table 3. Varied parameters for the take-over concept

Parameter	Description	Value
Tolerance band for lateral offset d_{lat}	Limits for deviation of the lateral offset between predicted (driver) and ideal trajectory	**Small**: ±0.5 m
		Large: 0.5 m in direction of obstacle, 1.75m in direction of left lane boundary
Dwell time t_{dwell}	Period of time in which the criteria Δd_{lat}, $\Delta\dot{\psi}$ and $\Delta\beta$ must be fulfilled simultaneously	**Short**: 0.15 s
		Long: 0.4 s

The concept described above defines whether and when the driver is allowed to regain the vehicle control and was the main focus of the take-over strategy in this study. However, the transition between the system-side and the driver-side part of the vehicle control was modeled with linear curve increasing from 0 % to 100 % for the driver's part and decreasing from 100 % to 0 % for the system-side part. The transition time was set constant to 0.3 s referring to [8].

3 Dynamic Vehicle Road Simulator (DVRS)

This section presents the driving simulator at the Niedersächsische Forschungszentrum Fahrzeugtechnik (NFF), where the subject study took place.

The Dynamic Vehicle Road Simulator (DVRS) consists of three main elements: the half-vehicle mock-up, the motion platform and the visualization. As the mock-up is based on a real VW Golf VII GTD that was cut in half, the DVRS offers a nearly unmodified driver's seat with the familiar elements such as steering wheel, gas and brake pedal, the cockpit with speedometer and tachometer and control elements like the blinker lever. This allows a high degree of immersion and a faster adaption for the test subject to the simulation environment. For the communication between the participant and the test leader during a running study drive a microphone and speakers are installed inside the mock-up.

Fig.8. Dynamic Vehicle Road Simulator (DVRS)

The mock-up is mounted on a HEXaDRIVE motion platform by Simtec Systems which consists of six hydraulic actuators. With this the platform is able to do both translational and rotational movements in all six degrees of freedom. This allows the DVRS to recreate the forces action on the driver during acceleration, deceleration and cornering. To visualize the created driving scenarios and therefore the driving environment seven displays were installed. Five of the are curved monitor that are mounted in front of the mock-up and cover an angle of nearly 180°. Two additional smaller displays are placed in the exterior mirrors on both sides.

On the software level IPG CarMaker was used to create the driving scenarios (4.1). The same dual-track model that was already used in 2.1 and 2.2 was implemented in the model environment of CarMaker4Simulink. The mock-up data like steering wheel angle, gas and brake pedal position were fed into the model via CAN-Bus communication to simulate the vehicle dynamics and the SbW evasion function respectively the take-over concepts. At the same time a corresponding steering feeling was generated with the built-in steering system based on a target steering torque generated within the SbW model and again transferred via CAN-Bus to the mock-up.

4 Study design

4.1 Take-over application variants

Within this study, four different applications of the take-over strategy (2.3) were investigated, that are described in the following **Table 4** and result from combinations of the parameters mentioned in **Table 3**. Also a reference variant V0, in which the driver had to perform the evasive maneuver by himself without any additional assistance, was included in the study procedure. Two parameters of the take-over model were varied: firstly, the width (and the shape) of the tolerance band that was defined on the basis of the ideal trajectory, and secondly, the time that the driver (respectively the predicted trajectory based on the drivers steering wheel input) must stay within the defined tolerance band, before receiving control of the vehicle. The transition time within which the control is transferred from the system to the driver remained constant 0.3 s.

Table 4. Definition and description of the investigated application variants of the take-over function

Application	Description
V0	**Manual** evasion (no take-over required): The driver defines the driven trajectory by his steering wheel input, there is no active steering intervention by the SbW system and therefore no take-over
V1	**Small tolerance band + shorter dwell time**: The predicted trajectory based on the drivers steering wheel input may deviate from the ideal trajectory by 0.5 m in both directions. The predicted trajectory has to remain within the boundaries for 0.15 s
V2	**Small tolerance band + longer dwell time**: The predicted trajectory may deviate from the ideal trajectory by 0.5 m in both directions. The predicted trajectory has to remain within the boundaries for 0.4 s
V3	**Large tolerance band + shorter dwell time**: The predicted trajectory may deviate from the ideal trajectory by 0.5 m in the direction to the obstacle and 1.75 m in the direction to the left lane boundary. The predicted trajectory has to remain within the boundaries for 0.15 s
V4	**Large tolerance band + longer dwell time**: The predicted trajectory may deviate from the ideal trajectory by 0.5 m in the direction to the obstacle and 1.75 m in the direction to the left lane boundary. The predicted trajectory has to remain within the boundaries for 0.4 s

In all application variants, also a steering torque was generated by the built-in steering system of the driving simulator, which supported the steering movement of the driver during the evasive maneuver. The steering torque was generated as soon as the evasion function was activated.

As there were four different applications plus the reference variant to be investigated, five different urban driving scenarios had to be designed in the CarMaker Software (**Fig.9**). After a practice drive, in which the participants had to get used to the driving simulator, in particular the motion platform and the simulated vehicle dynamics, the five scenarios followed with a survey part between two drives. In each scenario the autonomous evasion function was triggered by an obstacle and was to be overridden by the driver with one of the four take-over applications. The route was designed as an urban route with a total length of 8.5 km, on which the hazardous situations occurred in the form of road users such as vehicles or pedestrians suddenly crossing the road. The obstacles were triggered at various points along the route as soon as the ego-vehicle reached the LPTS for a vehicle speed of 50 km/h and a road friction coefficient of μ_R=0.6 (see 2.1, **Table 2**). To ensure a constant initial speed of 50 km/h the subjects were instructed to drive with activated cruise control through the whole scenario. Furthermore, the scenarios were designed in such a way that the driver were forced to override the evasion function since oncoming traffic was placed in the area of the obstacles.

Fig.9. Examples of the driving scenarios created in CarMaker with vehicles and pedestrians as moving obstacles

4.2 Questionnaire

Table 5 presents the main questionnaire that has been used for the main subject survey. These questions were asked after every scenario, which were each driven with a different application variant. As the study was conducted exclusively with normal drivers, a simple 5-point scale (1 to 5 for unipolar and -2 to +2 for bipolar questions) was used to answer the questions, except the first one, which was a simple Yes/No question regarding the perceptibility of the evasive function. In addition to the scale, each value was assigned to an attribute. The evaluation of these criteria follows in the next chapter.

Table 5. Questions for the subject survey after each scenario

Q1: Did you recognize any intervention during the evasive maneuver?				
Yes			No	
Q2: Did you have in control of the vehicle during the evasive maneuver?				
1	2	3	4	5
At no time	Barely	Occasionally	Mainly	At any time
Q3: How safe did you feel during the evasive maneuver?				
-2	-1	0	1	2
Very unsafe	Unsafe	Acceptable	Safe	Very safe
Q4: How did the steering wheel movement displayed by the SbW function influence your own steering behavior?				
-2	-1	0	1	2
Very negative	Negative	Neutral	Positive	Very positive
Q5: How do you rate the steering feeling regarding comfort?				
-2	-1	0	1	2
Very negative	Negative	Neutral	Positive	Very positive
Q6: How do your rate the evasive maneuver overall?				
-2	-1	0	1	2
Very negative	Negative	Neutral	Positive	Very positive

In addition to the questions of the main survey, **Table 6** shows a question of the closing survey that was asked to each participant when they finished all scenario runs. It's the question of acceptance and allows multiple answers. This question is also evaluated in the next chapter.

Table 6. Question of acceptance, which was asked in the closing survey

Q7: Which of the witnessed SbW function variant would you accept? (Multiple choice)				
V1	V2	V3	V4	None of these

5 Results of the study

This section presents the results of the driving simulator study regarding the questions Q1 to Q7 of the questionnaire in **Table 5** and **Table 6**. The data shown in **Fig.10-Fig.***12* are normalized using the z-standardization. In total there were $N=35$ participants between the ages of 18 to 49. The subjects were quite experienced drivers with driving experience between 6 and 20 years.

In **Fig.10** the evaluation of the questions Q1 and Q2 of the questionnaire are presented. As expected the majority of the test subjects recognized the intervention by the autonomous SbW function in the application variants V1 to V4, so most of the participants were aware that they were being assisted in the evasive maneuver. However, it also can be seen that the perceived intervention of the SbW function is also accompanied by a (more or less significant) reduction in the subjective feeling of controllability. While the reference variant is predominantly rated as controllable "at any time" (which was actually the case), the application variants with the shorter dwell time of 0.15s (V1 and V3) are rated as "mainly" controllable and the variants with a longer dwell time of 0.4s (V2 and V4) are even rated lower in the range to "occasionally" controllable. Since a longer dwell time means a longer period during which the driver is fully decoupled from the vehicle, this result seems quite plausible and may not necessarily be assessed as positive or negative.

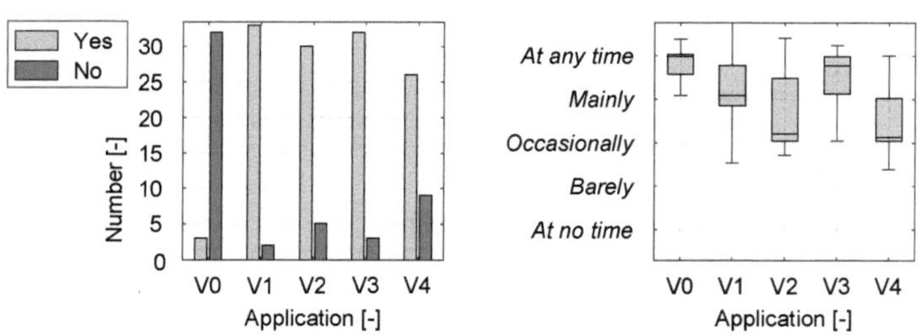

Fig.10. Results of Q1 (recognition of the steering intervention) and Q2 (controllability of the vehicle)

In the **Fig.11** the results of the questions Q3 and Q4 of the questionnaire are shown. Regarding the results of the previous question Q2 the influence of the "controllability" aspect on the subjective feeling of driving safety is recognizable in the left boxplot.

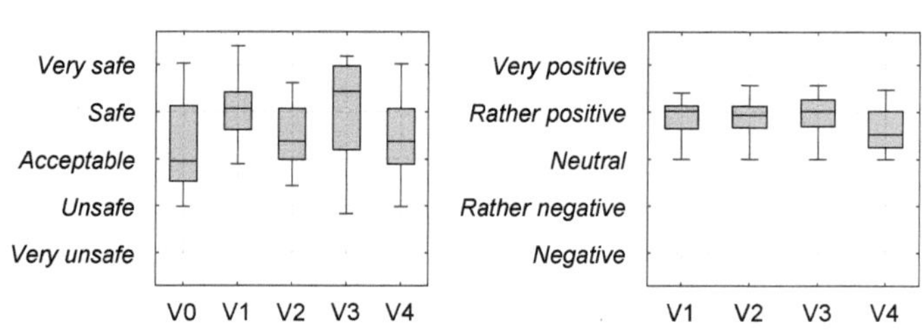

Fig.11. Results of Q3 (feeling of safety) and Q4 (influence by steering wheel movement)

All four application variants (even the reference variant V0) were at least rated with an "acceptable" safety feeling. While the variant V1 receives the rating of an "safe" and variant V3 even reaches a "very safe" feeling of driving safety, the two variants with the longer dwell times V2 and V4 clearly reduce the feeling of driving safety which could be very probably be related to the longer decoupling of the driver and therefore the lower rating regarding the vehicle controllability (Q2). Nevertheless, the median of V2 and V4 is still slightly higher than that of the reference variant V0, since the ideal evasion trajectory still assists the driver and ensures an earlier and less hectic evasive maneuver compared to the scenario where the driver needs to steer completely manual and may overreact.

Fig.*11* also shows a rather positive rating of how the steering wheel movement provided by the SbW system influenced the driver's reaction. It can be seen that there are no significant differences among the application variants since all variants provided the same feedback to the driver. All four were rated in the range of "neutral" to "rather positive" which indicates a slightly moving steering wheel may be a good haptic feedback for the driver to be made aware of the autonomous SbW intervention.

The **Fig.12** shows the results of the last two questions Q5 and Q6 of the questionnaire. First it can be seen there is once again a rather positive response from the subjects to the steering wheel feedback during the evasive maneuver and no significant differences between the four application variants. It also shows no (negative) influence of the take-over concept on the steering feeling when it comes to the transition phase where the driver cooperates with the SbW system for a short time. V1 to V4 are rated nearly the same as the reference V0.

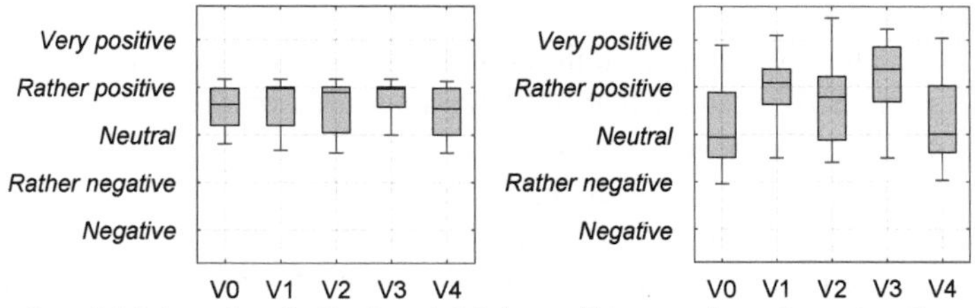

Fig.12. Results of Q5 (steering feeling) and Q6 (overall impression of evasion function)

Furthermore, in the right boxplot there is the overall rating of the driving scenarios. The reference variant has a larger spread since the result of the maneuver depended on the driver's reaction and driving skills to avoid the collision. So the reaction could result in a more or less critical situation which then would be rated as rather positive or negative. Regarding the median all four SbW application variants with take-over option were rated higher than the reference variant, which clarifies the potential of such a function. Looking at the differences between these variants it can be seen that the concept of a larger tolerance band (to give the driver more leeway) in combination with the shorter dwell time (to reduce the decoupling) performs best and tends to be rated as very positive. The variant with the smaller tolerance band (V1) and the same dwell time performs about as good as V3. The two variants with the longer dwell times were rated significantly lower whereby V4 even barely reaches a "neutral" rating (median) and is therefore about the same rating as the reference (manual) variant.

Lastly **Fig.13** shows the result of Q7, which is the one from the closing survey and which asks about the overall acceptance of the four application variants. Here it is clear to see, that the variants V2 and V4 are least accepted, while the V3 is the most popular variant and was chosen by 19 of 35 participants, followed by V1 (12 of 35 participants). In this question there was also the option for "none of these" if a participant would accept none of these variants at all, but this option was not voted once so it's not added to the bar chart.

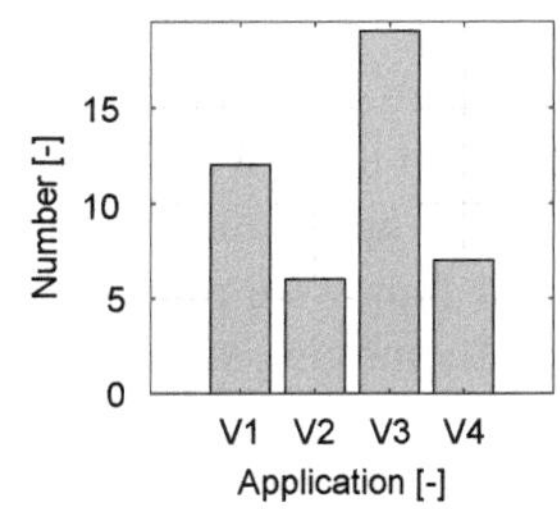

Fig.13. Result of the question of acceptance from the closing survey (Q7)

6 Summary

This paper reports on a driving simulator study that was performed to determine the acceptance of different take-over application variants of a Steer-by-Wire functionality. This paper focuses on the decision, if and when the driver is allowed to regain the full control of the lateral vehicle dynamics. For the study a dual track model with a validated model of a real SbW system, that has been measured on the institute's Hardware-in-the-Loop steering test bench, was used. A previously modeled SbW function in the form of an autonomous evasion function was the basis for the study and was then extended with a take-over concept for this purpose. The concept is based on the comparison between the ideal trajectory, that was driven by the SbW function autonomously, and a predicted driving condition which would result from the driver's steering wheel input. A set of parameters defining when the driver is allowed to regain the control after he was decoupled of the lateral vehicle movement, that were then varied to receive four different application variants of the take-over concept that were investigated in this study. The participants were asked on criteria such as the overall feeling of driving safety, the controllability and the steering feedback they received during the evasive maneuver. Finally, they answered the question of the overall acceptance of these witnessed application variants of the take-over concept. Within the study a variant without any intervention of the SbW function was also driven by the subjects to evaluate the differences and the potential of such a SbW function.

The results of the driving simulator study show that a shorter period of decoupling and therefore a smaller influence of the autonomous evasion function is desired by the majority of the participants. The application V3, where a larger, asymmetric tolerance band around the ideal trajectory was defined and the criteria for the predicted driving condition based on the driver's steering wheel input needs to remain within this tolerance band for 0.15s, was rated the best. The provided steering feedback in form of an assisting steering torque, was also rated as positive.

References

1. Vollrath, M., Briest, S., Schießl, C.: Ableitung von Anforderungen an Fahrerassistenzsysteme aus der Sicht der Verkehrssicherheit. Berichte der Bundesanstalt für Straßenwesen (BAST), Heft F60 (2006).
2. Dang, T., Desens, J., Franke, U., Gavrila, D.M., Schäfers, L., Ziegler, W.: Handbook of Intelligent Vehicles. Chapter: Steering and Evasion Assist. Springer Verlag. 1. Auflage (2012)
3. Eckert, A., Hartmann, B., Sevenich, M., Rieth Dr. P.E.: Emergency Steer & Brake Assist – A systematic approach for system integration of two complementary driver assistance systems. Continental AG. Paper Nr. 11-0111. (2011)
4. Winner, H., Hakuli, S., Lotz, F., Singer, C.: Handbuch Fahrerassistenzsysteme. Springer Vieweg Verlag. 3. Auflage (2015)
5. Ahrenhold, T., Achilles, F., Lausch, H., Henze, R.: Driving simulator study on the acceptance of a Steer-by-Wire steering system during braking on mue-split. Chassi Tech plus, Munich (2023)
6. Breuer, B., Bill, K.H.: Bremsenhandbuch. Springer Vieweg Verlag, 3. Auflage (2006)
7. Mitschke, M., Wallentowitz, H.: Dynamik der Kraftfahrzeuge. Springer Fachmedien GmbH, Wiesbaden (2014)
8. Hesse, T., Schieben, A., Heesen, M., Dziennus, Marc., Griesche, S., Köster, F.: Interaction design for automation initiated steering manoeuvres for collision avoidance. 6. Tagung Fahrerassistenzsysteme (2013)

Subjective and objective evaluation of steering feel on a steering HiL test bench

Alexander Haas1[0000-0001-5801-039X], Gregor Menze1[0009-0004-8768-5483], Christoph Bittner[1], Pascal von Kamp[1] and Dieter Schramm2[0000-0002-7945-1853]

[1] Dr. Ing. h.c. F. Porsche AG, Porschestraße 911, 71287 Weissach, Germany
[2] Department Mechanical Engineering, University of Duisburg-Essen, Forsthausweg 2, 47057 Duisburg, Germany
alexander.haas3@porsche.de

Abstract. Test benches represent a valid tool to transfer vehicle-based assessments to a laboratory environment allowing efficient testing in preliminary stages of the automotive development process. For steering systems, the achievement of a brand specific steering feel represents a fundamental goal of the development. Due to the subjective nature of the term "steering feel," objective metrics need to be identified, which accurately describe individual aspects of steering feel. In this contribution, a steering test bench it utilized to assess steering feel outside the vehicle. Therefore, fifteen hardware variants and sixteen software variants of one steering system are subjectively evaluated by an expert driver in the vehicle. For the objective description of these evaluations, a 4-Layer-Testspecification is derived, which ranges from ideal approximation of driving scenarios in layer one to characterization approaches of the passive steering system in layer four, aiming to study the influence of abstraction on the correlation results. Additionally, the test bench is expanded to a static driving simulator, where the surroundings are either displayed on a 2D monitor or virtual reality glasses. The results of the correlation analysis identify various objective metrics which successfully describe the selected subjective criteria. Driving simulator assessments also correlate with the reference evaluations from the vehicle. A final validation, where the derived linear correlation models are utilized to calculate the subjective assessments, demonstrates the good approximation. Consequently, the results of this contribution facilitate the utilization of a steering test bench for subjective evaluation of steering feel.

Keywords: steering feel, steering test bench, correlation analysis.

1 Introduction

As summarized in [1], the modern development process of passenger vehicles is confronted with reduced development time and less full vehicle prototypes while simultaneously encountering increased requirements by the customer and by law. As a subsystem of the vehicle, these challenges also apply to the development of steering systems. Nowadays, most vehicles are equipped with electric power steering (EPS) systems,

P. E. Pfeffer (Ed.), *15th International Munich Chassis Symposium 2024*,
Proceedings, https://doi.org/10.1007/978-3-662-71064-7_13

wherein an electric motor provides the necessary assistance forces to support the driver during steering maneuvers. These systems have replaced their predecessor, the hydraulic power steering (HPS), where the servo force is generated by the pressure difference of the hydraulic fluids in two adjacent chambers, controlled be the twisting of the torsion bar. One major advantage of the EPS is its power-on-demand principle, allowing a reduction in energy consumption leading to a fuel saving potential up to 0,8 l/100km compared to its predecessor [2]. EPS systems furthermore simplify the implementation of driver assistance functions [2], facilitate the application for different vehicle derivates [3] and provide improved packaging potentials due to the elimination of the hydraulic power supply [3]. Disadvantageous of EPS systems is the deterioration in feedback performance [4]. The term steering feedback represents the transmission of alterations in rolling resistance, lateral tire forces and wheel load to the driver in form of variations in steering torque [2]. Due to large, requested servo forces, exemplarily during parking, comparatively large electric motors are required. For sporty and high-performance vehicles, a steering design is implemented which utilizes two serial transmission stages, to translate the rotation of the electric motor into a translation of the steering rack [2]. This steering layout is called axle parallel EPS (EPSapa) since the electric motor is mounted parallel to the steering rack axis. A belt drive and a ball screw gear are utilized for transmission [2]. As a result of two transmission stages in combination with the rotor's and ball screw nut's inertia, an equivalent translational mass can be calculated [5]. depending on the steering systems, this mass reaches a magnitude of 1500 kg [6], leading to a low-pass filter behavior of the systems. The implementation of methods to encounter these shortcomings and the necessity of highly dynamic control approaches further increase the system complexity [5], tightening the challenges already originating from the demands of the development process.

To encounter these challenges early in the development process, the utilization of test benches gains in importance, since they allow early development without the requirement for vehicle prototypes. One goal of the steering system development is the achievement of a brand-specific steering feel. Steering feel represents the subjective perception of the dynamic behavior of the vehicle caused by a steering wheel command of the driver [2]. Additionally, it is divided into steering- and response behavior [2]. Here, the subjective, vehicle-based evaluation represents the most realistic and meaningful approach and still functions as the benchmark for all other methods. Trained test drivers with detailed knowledge and understanding of the goals for the brand specific realization of steering feel are utilized for these purposes [7].

To generate a linkage between the subjective perception and objective metrics numerous studies aim to identify these connections on vehicle level. Based on these activities different normed driving maneuvers like the Weave-Test in ISO 13674-1 [8] and the Transition-Test in ISO 13674-2 [9] have been derived. These examples represent open-loop driving maneuvers where the steering wheel angle, vehicle speed and other aspects are predefined aiming to minimize the influence of the driver [10]. Exemplarily, [11] investigated a driving maneuver similar to the Weave-Test procedure and derived characteristic values based on two-dimensional plots of steering wheel angle, torque, lateral acceleration and steering work gradient to describe on-center behavior.

Based on similar investigations with a correlations analysis, [12] identified the Weave-Test maneuver as the most relevant for On-Center assessment.

Besides open loop approaches, there also exist closed loop maneuvers. The most known closed-loop vehicle test, where a driver is allowed to incorporate the vehicle response [10], is the double lane change maneuver from ISO 3888-1 [13].

In [14] parameters from open-loop maneuvers were utilized to describe the subjective assessment from a double lane change maneuver. Although the open-loop maneuvers provided relevant correlations, metrics directly derived from the double lane change maneuver are more relevant. Frequency domain-based correlation investigation was performed in [15] wherein an analytical form of the vehicle transfer function for the double lane change maneuver was derived. Display of yaw amplification over the identified time delay in a Weir-di Marco diagram led to the identification of a positively perceived goal range in the diagram. Additional frequency domain investigations were performed in [16] and [17].

A large investigation with the subjective and objective assessment of 25 vehicles was performed in [10]. Due to the considerable number of tested vehicles, which originated from five different segments, not only relevant correlations, but also an indication of vehicle segment-dependencies were derived.

Vehicle based investigation up to the vehicle limit were conducted in [18], where subjective assessments of vehicle limit characteristics were correlated to objective data from open-loop maneuvers. In [19] changes in road friction and their feedback indication to the driver was studied. In other investigations on steering feedback, [20] and [21] identified the acceleration at the steering wheel as an objective measure to describe the feedback behavior of the steering system. Faults of the steering system and their perception by the driver were analyzed in [22]. Based on the enveloping curves of the resulting steering and vehicle reaction signals, correlations analysis was conducted.

In contrast to vehicle-based assessments, steering test benches increase flexibility and allow the investigation of the real steering system outside the vehicle. Therefore, different publications study steering systems on test bench level.

Investigations on a steering test bench for on-center steering behavior were conducted in [23]. Here, tests with active servo assistance as well as inactive assistance are used to generate objective data. Based on subjective assessments by test drivers, correlation analysis is performed, and valid correlations are identified. For steering feedback evaluation [24] conducted a study to correlate subjective assessment from the vehicle with metrics derived on a test bench. It also emphasized the added value of the incorporation of a human grip model on the correlation results. Other approaches on steering test benches focus on the characterization of steering systems [5,25], exemplarily by identifying the transfer function of a steering system, without presenting a clear correlation to subjective assessments.

Similar to steering test benches, driving simulators can be implemented to subjectively perceive a steering system outside the vehicle. For these cases, either force-feedback steering wheels, or real steering system can be used for subjective assessment. Exemplarily, [22] utilized a steering test bench, which was expanded to a driving simulator with the real steering system, to assess occurring steering errors. On a similar setup [26] implemented a driving simulator to improve the application of a steering function which correlates to the subjective assessment in the vehicle.

Based on the presented investigations on vehicle level, it can be stated that both time-domain and frequency domain approaches deliver valid correlations between objective metrics and their subjective assessment. Furthermore, objective metrics from driving scenarios closely related to those utilized for subjective assessment tend to exhibit better correlations than generic driving maneuvers.

On steering test benches, the real steering system with its imperfections, variations and the real steering control unit is available. Additionally, a steering test bench can be expanded to a stationary driving simulator with the real steering gear to directly assess the steering system subjectively. For these reasons, the goal of this paper is to further improve the utilization of steering test benches for the development of steering feel within the automotive development process by answering the following questions:

1. Is it possible to subjectively assess steering systems on a steering test bench expanded to a driving simulator and what role does the visualization play?
2. Which objective metrics, derived on a steering test bench, are relevant for steering feel?
3. What level of accuracy is required to derive valid correlations?

The objective metric to compare the quality of correlation is the Pearson coefficient r with its definition [23]

$$r = \frac{\sum_{i=1}^{n}(x_i-\bar{x})\cdot(y_i-\bar{y})}{\sqrt{\sum_{i=1}^{n}(x_i-\bar{x})^2\cdot\sum_{i=1}^{n}(y_i-\bar{y})^2}}, \tag{1}$$

The Pearson coefficient lays in the range between [-1, 1] and describes the extent to which the sample x_i, here the objective metric, with its mean value $\bar{x}$ correlate linearly with the sample y_i, here the subjective assessment, with its mean value $\bar{y}$. The magnitude describes the linearity while the sign incorporates the direction of correlation.

To identify these answers, the residual of this paper structures as follows: chapter two describes the necessary preparations for correlation analysis, including the derivation of a subjective criteria classification scheme, the implementation of a 4-Layer-Testspecification, literature review on known correlations and the generation of steering system alterations. The following section describes both the final questionnaire and the subjective evaluation procedure. In chapter 4 the Hardware-in-the-Loop (HiL) test bench and the vehicle model implementation are described while, chapter 5 presents the derived results. A summary and outlook conclude this paper in chapter 6.

2 Preparation

For the correlation analysis, subjective criteria, steering system variants and a test specification are required. This section elaborates these preparations including a categorization of known objective data.

2.1 Derivation of Subjective Criteria

In order to adequately assess the steering feel of a steering system, the selection of important and precise criteria for its description is fundamental. Although each original equipment manufacturer (OEM) possesses their individual criteria, they are normally not openly accessible and there is no clear agreement on the correct set of criteria among various publications [27]. One approach which is commonly used for the classification of subjective criteria is a layer model, as exemplarily described in [28]. Based on different abstraction levels, steering feel is refined until the most basic description or criterion is found. Similar approaches are used in [10,12,29]. Due to the easy visualization of criteria in such a structure, its possibility to be further expanded and its repeated implementation in academic work, a similar setup is incorporated. The process for the identification of relevant subjective criteria is displayed in Table 1 and contains literature review, reduction, renaming and clustering of the criteria.

Table 1: Process for the identification and classification of subjective criteria

Step #	Content	Detailing
1	Literature review and collection of subjective criteria describing lateral vehicle behavior	~ 430 criteria from ~ 30 references e.g. [10,30,31]
2	Elimination of duplicating and irrelevant criteria	Exemplarily criteria which are not influenceable by steering system
3	Generation of mutual understanding and optional renaming	Agreement on definition
4	Generation of a reference tree structure	Consideration of boundary conditions and focus on generic structure
5	Classification of the criteria in the structure	Localization of the criteria within the derived structure.

The described process is conducted in close cooperation with an expert driver with more than ten years of experience in the subjective assessment of vehicles. Following this collaboration, it is ensured that all relevant criteria are incorporated and that both parties understand the resulting structure. Fig. 1 displays the derived structure for one exemplary path through all the abstraction levels. Level 2 and level 3 of this structure are primarily utilized to refine the surrounding conditions, namely velocity ranges and the driving situation of the vehicle. On level 4, the same four categories are applicable for each combination of level 2 and level 3 attributes. For "Correction Effort," the focus of investigation is set to very small excitations by the driver to the steering wheel and how the vehicle responses. "Steering Effort" deals with rougher excitations by the driver, meaning larger steering wheel angles and torques. The third category is displayed as "Haptic Perception." Here, discontinuities, friction and damping are relevant. "Feedback" incorporates all information transmitted from the road or steering system to the driver. Classical approaches, as summarized in [6], utilize frequency ranges to categorizes feedback as useful or not useful. Within this contribution, feedback is regarded as useful if it provides support to manage the current driving situation. Exemplary, feedback due to wheel imbalance is typically regarded as not useful feedback. In the definition within this paper, it is useful, since it indicates a shortcoming of the vehicle, but only until the resulting oscillations disturb the driver in its task. The herein displayed example further structures the feedback during cornering into three categories in level 5, where "Road Surface Condition" is further separated into two criteria on level 6. These criteria represent the most refined aspects in this path and are utilized for the evaluation.

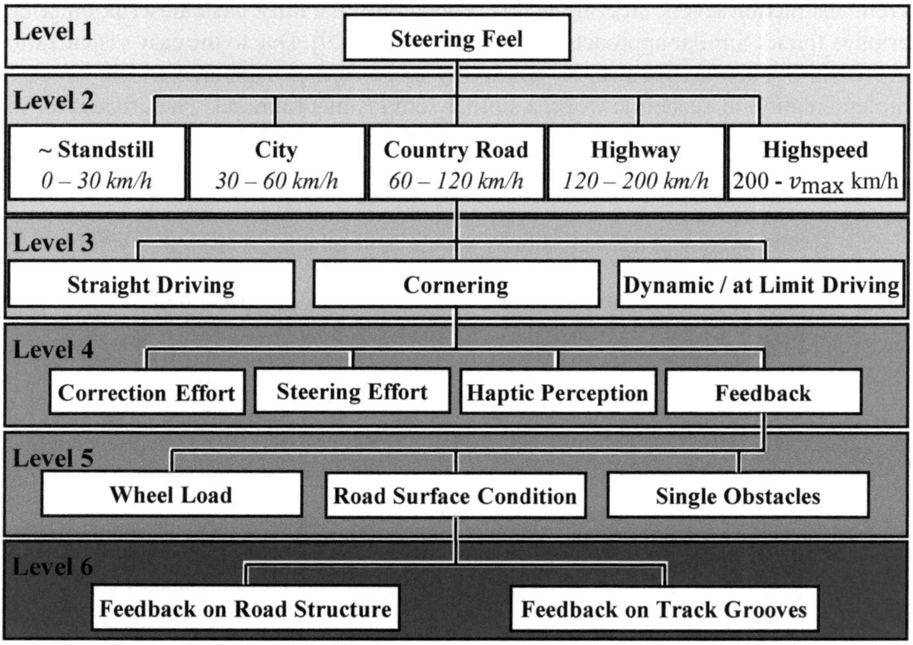

Fig. 1: Exemplary classification on six levels of refinement

For the investigations in this paper, a reduced questionnaire is chosen from the presented structuring. The details of the criteria, including definitions of level 6 aspect is given in subsection 3.1.

2.2 Derivation of the 4-Layer-Testspecification

To answer the third question in the problem statement, namely how detailed a driving situation must be modeled to derive relevant correlations, a 4-Layer-Test specification is implemented. The main goal of this approach is the reduction of modeling accuracy as the layer number increases. Simultaneously, the simplicity of these higher layers increases. Fig. 2 visualizes the basic concept behind each layer.

As displayed, the first layer aims to approximate the real-world conditions as closely as possible. Therefore, a single-track-model (STM), derived in subsection 4.2, is parameterized with measured data from the real vehicle.

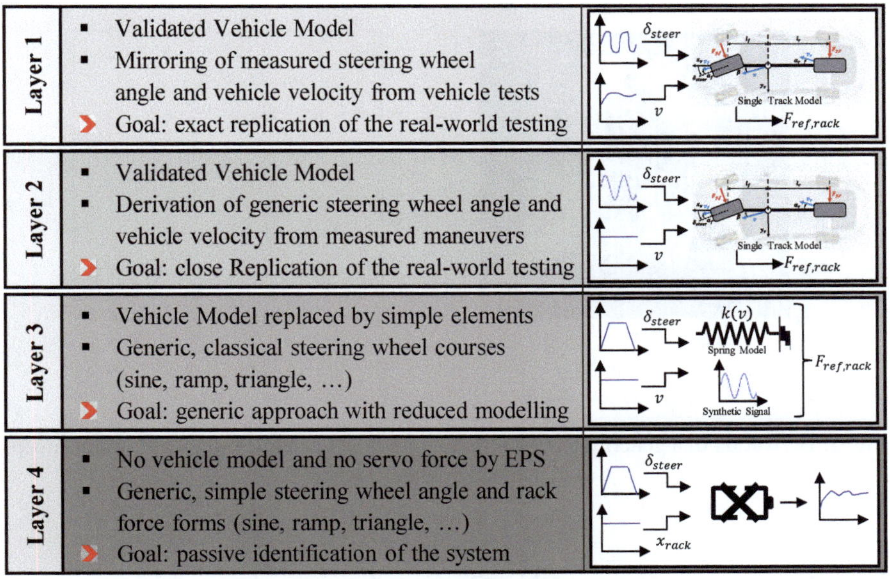

Fig. 2: Visualization of the 4-Layer-Test specification approach

For the second layer, the same vehicle model is utilized. In contrast to layer one, both steering wheel angle (SWA) and velocity reference signal are transferred to a generic form for each specific driving scenario. Therefore, an approach in accordance with [23] was followed. Based on multiple measurements of the same, relevant, driving scenario, a heatmap is constructed, see Fig. 3 (a), allowing the derivation of an average steering wheel angle course Fig. 3 (b).

The third layer is not linked to the vehicle model. Instead, a virtual spring is implemented, which reflects the behavior of the vehicle in a linear driving range. Additionally, generic reference force signals are generated to investigate the response of the steering system to sine, Dirac and jump excitations. Also, the steering wheel angle signal is now reduced to basic forms like sine, ramp, and triangle. Based on the target lateral accelerations $a_{y,target}$ of the individual scenario, the rack force signals can be approximated by the knowledge of the vehicle mass m_{veh}, the kinematic τ_k and pneumatic trail τ_p, the steering linkage ratio i_{SLR} and the assumption of a centered center of gravity [32]:

$$F_{rack} \approx \frac{m_{veh} \cdot a_{y,target} \cdot (\tau_k + \tau_p)}{2 \cdot i_{SLR}} \tag{2}$$

On the lowest layer, the fourth, there is no steering assistance. Here, only passive maneuvers are performed, which aim to analyze the basic mechanics of the system. The tests now focus on basic investigations with generic excitations. Although these tests exhibit the largest deviation from the tests on vehicle level, they are applicable to any steering system, at any state since no assistance force is required. Therefore, investigations on the fourth level are especially relevant.

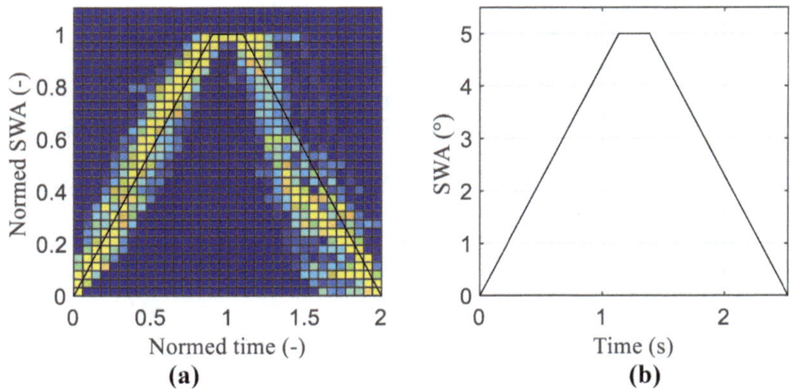

Fig. 3: Derivation of a generic steering wheel angle (SWA) course from a normed heatmap (a) to the time domain signal (b)

2.3 Classification of known Objective Correlations

There exist several reported objective parameters, which correlate with subjective aspects of the lateral dynamics of a vehicle. A summary of the individual metrics is beyond the scope of this paper, but based on a literature review, distinct groups regarding the representation of data can be derived. The following summarized categories are all relevant for the herein investigated tests.

Time domain metrics of a single signal

This category describes classical characteristics of a single signal for a given driving maneuver. Examples for these metrics are the maximum steering torque during step-input test [33] or steering torque differences during road friction jumps [19]. These metrics are only valid for the investigated scenario.

Time domain metrics of multiple signals

Similar to the previous category, the evaluation here takes place in the time domain and incorporates the time delay between two signals, as exemplarily the time delay between steering wheel angle and lateral acceleration or yaw velocity in [34].

Two-dimensional representation

For this category, two measured signals are plotted on the x- and y-axis of a Fig.. One of the most know example is the visualization of the steering torque over the steering wheel angle during the weave test [8]. In this category, gradients, zero crossings, y-values at defined x-values and hysteresis widths are relevant [8].

Frequency domain parameters

Aside from time domain based approaches, different driving scenarios, exemplarily the frequency response test [35], are applicable for frequency domain evaluations. This contains phase delays between two signals [33] and their transfer function gain [10].

2.4 Derivation of Steering system variations

Since a range of subjective ratings for different criteria is needed to identify relevant correlations, both software and hardware alterations are implemented on the steering systems. For this purpose, a total number of thirty-one steering system variations are derived, where fifteen variants inherit changes on the hardware and sixteen variants are implemented by changes in software and application. The following Table 2 summarizes the applied changes to the steering systems:

Table 2: Summary of the implemented variations for the subjective evaluation in the vehicle used for correlation analysis

Hardware Alterations		**Software Alterations**	
Descriptions	**Number of variations**	**Descriptions**	**Number of variations**
Yoke Preload	11	Dynamics Adaption	9
Friction Levels	2	Feedback Module	2
Steering Ratio	2	New Module for feedback increase	7
Bellow mounted	2		
Pinion Sealing	2		

It is worth mentioning, that the columns do not add up to the previously described number of hardware and software alterations. This is caused by the utilization of the reference steering system in each row. Therefore, the number of columns must be subtracted from the sum of all columns and added by one to get the correct number of variants. Although most changes can be implemented by alterations on one steering gear, for friction level and steering ratio, additional steering systems are required. Therefore, a total of three steering systems EPS1 with high friction, EPS2 with low friction and EPS3 with a different steering ratio are utilized. To investigate only one difference, the goal to apply only one alteration at the same time is followed. This goal is neglected, exemplarily to reduce the number of steering system conversions in the workshop for a more efficient process. An overview on the basic procedure of application derivation ca be found in [24].

3 Subjective Assessment

The key aspect of this contribution is the derivation of subjective assessments from vehicle test. Here, the utilized questionnaire and the evaluation procedure are explained.

3.1 Questionnaire for the evaluation of steering feel

Table 3 summarizes the relevant subjective criteria including their description.

Table 3: Overview of the considered steering feel criteria for straight driving

#	Nomenclature	Description
1	Straight, Response Behavior	This criterion describes the correlation between the smallest steering input variances and the corresponding vehicle reaction during straight driving.
2	Straight, Friction Impression	This criterion describes the perceived, constant and speed-independent counterforce during steering maneuvers with low steering angle speed or during steering response maneuvers during straight driving.
3	Straight, Feedback on Single Obstacles	This criterion describes the feedback of individual road events such as transverse edges and potholes to the driver when driving straight ahead. This also includes individual stimuli that stand out from the general stimuli of the road structure.
4	Straight, Feedback on Road Structure	This criterion describes the feedback of the road surface condition when driving straight ahead. It evaluates how clearly smooth and rough surfaces are perceived, whereby the relationship between the feedback and the vehicle reaction is important.

Additional criteria 5 - 8 are regarded, which exhibit an identical definition but are assessed during cornering. For the questionnaire, it is also necessary to provide an adequate rating scale to allow easy assessment by the evaluator while maintaining as much information as possible. A well-known rating scale within the automotive industry is represented by the ten-mark-scale (TMS) [7]. Here, "10" represents the currently optimal implementation, "5" represents the lowest rating still regarded as industry standard and "1" is the worst performance. Disadvantageous of this scale is the missing information in which direction the optimum is located. Exemplarily, if road structure feedback is rated with a "7", it is not accessible, whether the feedback is perceived too little or too much. This can be compensated by implementing an additional "direction of deviation" (DOD) scale, as exemplarily utilized in [10]. Here, three additional boxes are provided ranging from "too little" to "too much" while the middle represents the ideal implementation.

It is worth mentioning, that the herein utilized scale is an absolute scale, since there is no direct connection to a reference setup. Problematic with this approach is the finding in [36], that relative rating improves the correlation results. Therefore, a reference setup is defined for the assessment and all subsequent variants are rated relative to the reference setup on the provided absolute scale. As rating step size, 0.5 is selected. An extract of the questionnaire, containing the classification information as well as the rating system, is displayed in Fig. 4.

The usage of the subjective assessment for subsequent correlation analyses, requires the fusion of the TMS and the DOD to one value. Based on the deviation direction, the TMS is either multiplied by "-1" if the implementation is "too little", or by "+1" if it is

#	Level 4	Level 5	Level 6	TMS	DOD
Level 2 – Velocity Range 60 – 120 km/h					
Level 3 – Straight Driving					
4	Feedback	Road Surface Condition	Feedback on road structure	☐	Too little ☐■☐ Too much

Fig. 4: Extract of the subjective questionnaire for subjective criterion number 4

"too much". At this point, the continuity is lost, since a TMS of 9 is closer to an ideal "10", than a rating of "- 9". Therefore, in accordance with [10], the scale is transferred into a range from " - 8" to "+8", where "0" represents the optimum, called the transformed scale (TF). This approach orders the ratings and displays identical deviations from the ideal mark for each TMS, regardless of the DOD rating. The transformation logic shows in Fig. 5.

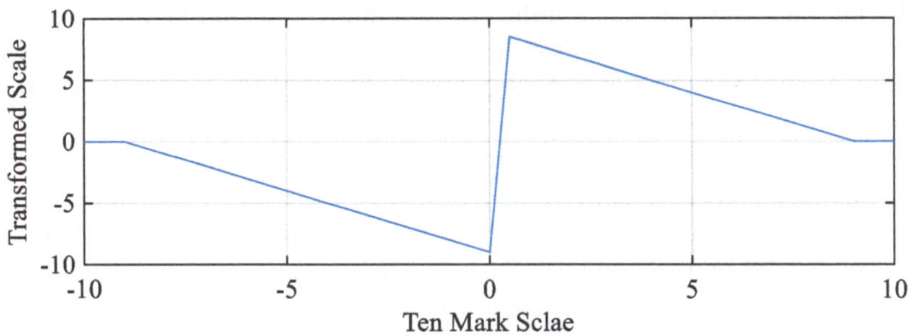

Fig. 5: Transformation lookup table from the ten mark scale to the transformed scale

3.2 Subjective Evaluation Procedure

For the subjective evaluation of the described aspects of steering feel, all assessments are performed by one experienced expert test driver who was also involved in the derivation of the questionnaire. Advantageous of this approach is the common understanding of the criteria by both the assessor and the instructor. Additionally, as summarized in [37], it was stated in previous research that a driver dependence in the assessment can be found, even if expert drivers are utilized. By regarding only one assessor, these deviations can be mitigated and need no further statistical compensation, impeding problems for correlation investigations.

Other preconditions like the utilization of the same proving ground for all assessments, dry weather at +15 °C ± 5 K, identical vehicle and tire conditions were considered during the investigations. By utilization of an electric steering column and seat, it was also possible to assure identical positioning of the driver in the vehicle for all setups. The vehicle for investigation represents a sport-vehicle with high demands for feedback and steering dynamics.

Before the assessment of the first steering variant, which represents the reference steering system for all subsequent evaluations, the assessor was instructed to rate each criterion as independent as possible. It was emphasized, that the steering variants are not final products but aim to adjust different characteristics individually. Therefore, a depravation in one criterion does not necessarily lead to an overall decrease in performance. Also, a possibility was given to clarify any open points before starting with the evaluation.

For the first steering system variation, which is communicated as the reference setup, the assessor was given the chance to adjust to the vehicle, develop a good understanding of the steering performance and accommodate with the questionnaire. Additionally, the evaluator freely selected specific driving scenarios which allowed the evaluation of each criterion. Consequently, each item on the questionnaire was linked to a location on the proving ground and connected to a vehicle velocity. The requirement for the definition of the driving situation used for evaluation originates from two aspects: first, a clear connection between driving maneuver and subjective criterion simplifies the evaluation on the steering test bench. Second, by limiting the evaluation of a criterion to one driving state, environmental influences are limited to a minimum. This is especially important, since an assessor is only able to evaluate characteristics of a system, if they are perceivable [37]. Exemplarily, if the evaluation location of road structure feedback changes around the proving ground, it is not perceivable on smooth surfaces, but rather dominant on rough roads. Therefore, the given assessment may not be comparable if this variation is present.

After the first steering system, there is still no restriction on evaluation time. After communicating that a new steering variant is now implemented, the assessor started the evaluation procedure. After approximately five minutes, when all driving states have been experienced, the instructor guided through the questionnaire, controlling the compliance of the driving situations and the criteria. At the end of the questionnaire, a pause of approximately ten minutes was taken to counteract fatigue of the assessor. On average, the evaluation time took 20 minutes per steering variant. Due to mechanical changes in the workshop, there were also prolonged pauses to relieve the evaluator. During assessment, no information on the current steering variation was provided to the evaluator and variations were implemented in random order. Still, due to necessary, longer breaks, it cannot be excluded, that the evaluator expected hardware changes or for the shorter breaks, that only software was adapted. All three steering systems with all variants were evaluated within a range of three weeks, reflecting a compromise between the necessary workshop tasks on the one hand and the need for a short timespan to keep the reference setup in head on the other hand.

4 HiL Steering System Test Bench

The main component of this paper focuses on the improved utilization of a steering test bench within the development process. This subsection gives an overview of the investigates test bench and describes how the driving scenarios are implemented and utilized for the generation of objective data.

4.1 Overview

In the herein presented investigation, an electro-mechanic HiL steering test bench is utilized to analyze the steering system. A description of the test bench can be found in [38]. It consists of two permanent magnet linear synchronous motors (PMLSM), which can be connected to both sides of the steering rack. They are named in accordance with their mounting side in the direction of travel, namely rod right actuator (RRA) and rod left actuator (RLA). At the driver side, a steering wheel actuator (SWA) is implemented to introduce steering wheel excitations into the system. Additionally, a hardware-in-the-loop computer is part of the test bench, which is responsible for the residual bus simulation, the steering test bench control algorithms, and the maneuver definition. This also includes the possibility to load a vehicle dynamics model into the real time computer. The setup of the test bench is displayed in Fig. 6.

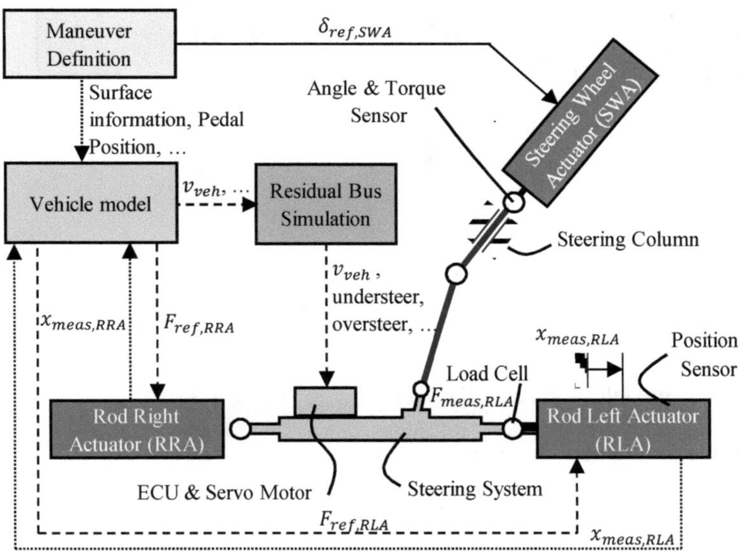

Fig. 6: Setup of the HiL-Steering-Test-Bench [39]

Here, only one PMLSM is connected to the steering rod. This can be justified by the sufficient excitation force of only one PMLSM and additionally increases the control performance if vibration mitigation strategies as in [40] are implemented. Within the maneuver definition, depending on the setup, reference steering wheel angle $\delta_{ref,SWA}$ and, in case a vehicle model is deployed, surrounding conditions are defined. Based on the measured rack displacement $x_{meas,RLA}$, which is introduced by the SWA, the vehicle model calculates the resulting rack force, which in term is utilized as the reference force signal $F_{ref,RLA}$ which implementation is measured by a load cell. The vehicle velocity and other relevant bus signals are provided in accordance with the state of the vehicle. Similar explanations and additional information on the steering test bench can be found in [24,39,40].

4.2 Model Derivation

Starting point for the reproduction of the investigated driving scenarios during subjective assessment is a single track model, as exemplarily described in [41,42]. The idea here is to shift the center of gravity down on the road surface, therefore hindering any influence of the vehicles inertia to change the wheel load distribution. Additionally, the two wheels on each axle are combined into one wheel. The basic concept of the STM is visualized in Fig. 7.

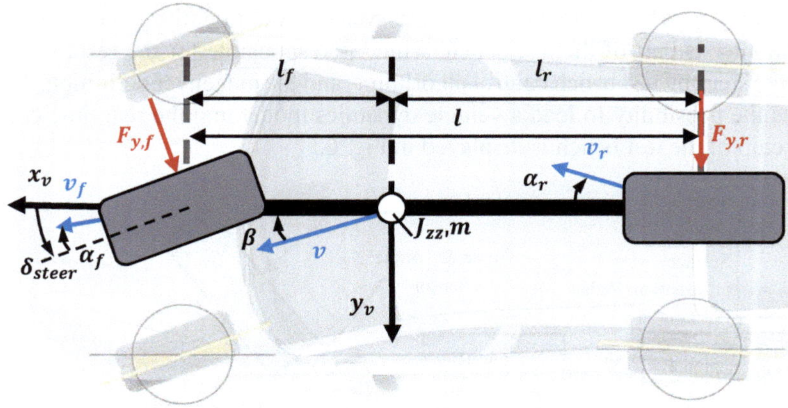

Fig. 7: Sketch of a single track model in reference to [41]

The impulse theorem in lateral direction is given as [41]

$$m \cdot (\dot{\psi}_v + \dot{\beta}) \cos(\beta) = \cos(\delta_{steer}) \cdot F_{y,f} + F_{y,r} \tag{3}$$

and for the momentum theorem

$$J_{zz} \cdot \ddot{\psi}_v = F_{y,f} \cdot \cos(\delta_{steer}) \cdot l_f - F_{y,r} \cdot l_r \tag{4}$$

Here m stands for the vehicle mass and $F_{y,f}$ and $F_{y,r}$ are the lateral tire forces at the front and rear, respectively. l_f is the front length and l_r the rear length of the vehicle while J_{zz} represents the rotational inertia of the vehicle. δ_{steer} represents the wheel angle, β the slip angle, ψ the yaw angle and v, v_r and v_f the velocities of the vehicle in the center of gravity, the rear and front wheel, respectively.

For the tire run-in behavior, a PT1 approach similar to [3] in the form

$$\frac{x_{f,r}}{v} \dot{F}_{y,f,r} + F_{y,f,r} = F_{y,f,r,stat} \tag{5}$$

is implemented, where $x_{f,r}$ is the run-in length of the front or rear tire, while $F_{y,f,r,stat}$ is the stationary tire force at the front and rear tire.

In the first step, linear tire skew stiffness $c_{\alpha,f,r}$ is used in the form of [41]:

$$F_{y,f,r,stat} = \alpha_{f,r} \cdot c_{\alpha,f,r} \tag{6}$$

where $\alpha_{f,r}$ is the tire slip angle at the front and rear.

In the second stage, a nonlinear extension is introduced by incorporating a Pacejka tire model, exemplarily given for the front tire, in the form [43]:

$$F_{y,f,stat} = Dsin(Carctan\{B\alpha_f - E(B\alpha_f - \arctan(B\alpha_f))\}) \tag{7}$$

allowing the expansion of the single-track model up to at limit driving.

For the parameter identification process, an on-center steering maneuver measured in the investigated vehicle, is utilized to estimate the linear tire skew stiffnesses and the run-in length at both axes. Information like vehicle mass, length etc. are taken from vehicle fact sheets. In a second step, the residual Pacejka parameters are estimated for a slow ramp steer maneuver up to maximum lateral acceleration. Both identification approaches utilize the MATLAB function fmincon [44].

A comparison of both maneuvers including the simulated courses are displayed for the on-center maneuver in Fig. 8 (a) and for slow ramp steer to limit in Fig. 8 (b).

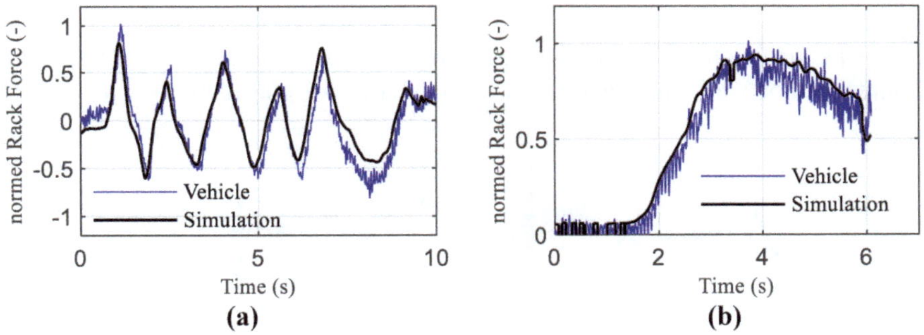

Fig. 8: Comparison of the simulated and measured rack forces for (a) on-center steering and (b) slow steer ramp into at limit driving

Now, the STM is capable of approximating both on-center applications as well as maneuvers towards the vehicle handling limit.

In order to analyze single obstacle crossings during cornering, the model is expanded by a more detailed approximation of the contact point of the tire, see Fig. 9.

Here, the tire is modeled as a spring without damping characteristics, which can be neglected due to the damping between m_z, the vertical vehicle mass, and m_w, the wheel mass, as motivated in [42]. The displayed stiffnesses and damping operate in vertical $(.)_z$ and longitudinal $(.)_x$ direction. During riding on a smooth surface, see Fig. 9 (a), no additional excitation is introduced. By implementing a lookahead window and the definition of the road geometry as table data, the shortest contact length from the wheel center to the road surface is calculated. In case contact to an obstacle occurs, a second spring goes into contact with the point in closest proximity to the wheel center, see Fig. 9 (b). The occurring additional force then leads to an impulse like force excitation. The resulting tire spring forces are translated by the wheel load arm r_{WLA} and interfering force arm r_{IFA} into a torque around the wheel axis. Then, the steering linkage gear ratio translates the resulting torque into a rack force. Similar to the STM,

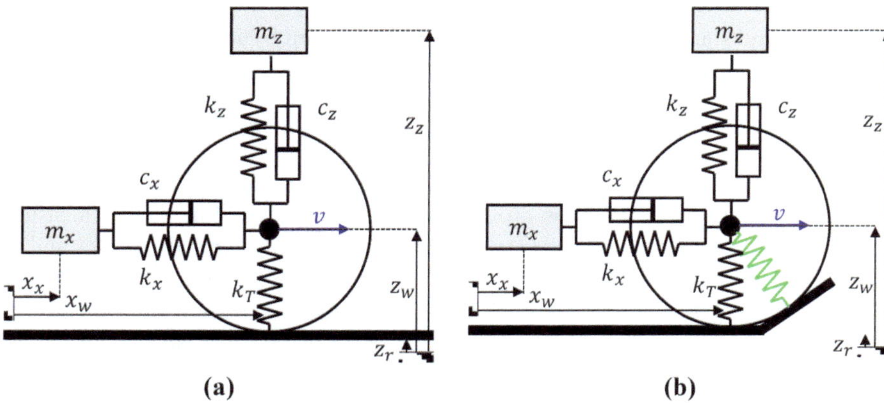

Fig. 9: Model structure of single obstacle crossing (a) for driving on smooth roads and (b) when in contact with the single obstacle

the MATLAB function fmincon is implemented for parameter estimation [44]. The good agreement between simulated and measured date is displayed in Fig. 10.

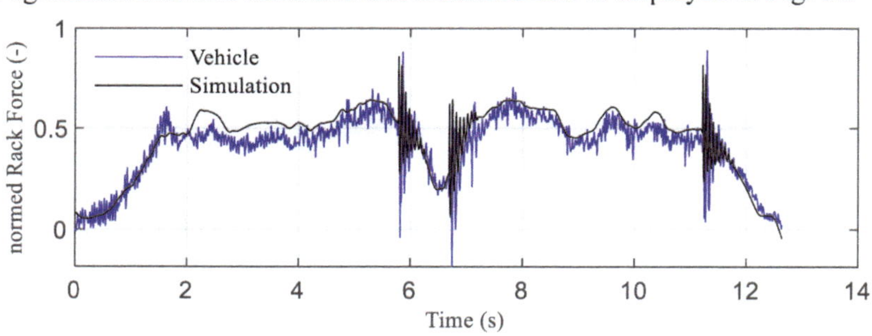

Fig. 10: Comparison of the measured and simulated rack force course for a single obstacle crossing

The last driving situation which requires modeling is the investigation of road structure feedback. Due to the fine structure of the road surface, a single contact model is no longer valid [42]. Since detailed tire models incorporate a large effort to be implemented, the excitations are introduced by information from measured power spectral densities (PSD) from the driving scenario within the vehicle. This approach is therefore only valid for the exact vehicle, but it represents a quick and accurate possibility to investigate these situations. Whereas the described vehicle model is implemented for the evaluations on layer one and two of the 4-Layer-Test specification, scenarios in the other two layers are derived from linear vehicle behavior by implementing a linear spring or predefined rack force courses.

4.3 Subjective Evaluation – Driving Simulator

For the subjective evaluation of the steering variants on the steering test bench, an extension of the existing setup to a static driving simulator was implemented. In this setup, the driver replaces the SWA and initiates a steering wheel input which is encountered by a steering rack force, originating from the vehicle model.

The assessor at the test bench is identical to the one from the vehicle evaluation procedure. He also exhibits some experience with driving simulations from video games, facilitating the acclimatization to the boundary conditions. In advance to the evaluation, it was again stressed that, each criterion should be assessed individually neglecting other shortcomings. Due to an expected difficulty to correctly assess the TMS value on the first try, since no experience on subjective steering feel assessment on a driving simulator existed, the reference system from the vehicle assessment was again utilized as the reference setup in the simulator. The evaluator was still instructed to grade the criteria like in the real vehicle. All other steering variants are rated in reference to the first setup.

In order to investigate the influence of the immersion technique, two visualization approaches were followed. First, a 2D curved monitor is installed for the visualization of the proving ground which is the same test track known from vehicle assessment. Second, virtual reality (VR) glasses were implemented to allow a 3D impression, including the interior of the vehicle. Both visualization approaches utilize an in-house solution.

The process of subjective assessment is directly linked to the vehicle evaluation. This is possible since the same proving ground is utilized, allowing identical boundary conditions. Depending on the current position on the virtual track, different excitations are triggered automatically or can be switched on manually, so each evaluation scenario stays identical.

Initially planned was the reevaluation of all steering variants, but due to nausea of the assessor, only the variants with boundary subjective rating were utilized. For VR glasses, nausea further increased which led to an additional reduction of the subjective criteria and number of steering system variants. For VR, only steering feedback evaluation is possible. During feedback evaluation, yaw behavior of the vehicle was quasistatic, reducing the driver's nausea. In conclusion, a reduced number of seven steering variants, two hardware and six software changes, was evaluated with the 2D monitor. For VR, five variants, two hardware and four software changes were implemented.

4.4 Objective Evaluation

The objective evaluation does not contain any instructions since the investigated maneuvers are predefined and implemented in an automated procedure, without huma influence. For the first two layers of the 4-Layer-Test specification, one test case is directly linked to one subjective criterion as the goal here is to adequately imitate the driving scenario utilized for subjective assessment in the vehicle. Therefore, two on-center maneuvers are regarded for response behavior and friction impression during straight driving. A similar signal with steering angle offset is selected for the cornering investigation. While an excitation signal for single obstacle crossing in straight driving is extracted from a measured PSD, the introduced obstacle model is utilized for cornering investigation. Road structure excitation also originates from a PSD for both straight driving and cornering.

For layer three and four, since the test cases become more generic, multiple testcases may contain information for different subjective criteria. To reduce the investigations to the most basic tests, layer three uses transfer function estimation techniques to derive the forward transfer path, meaning from the steering wheel to the rack, as well as the feedback path, from the rack to the steering wheel.

For the forward path, a steering wheel angle is predefined and implemented into the maneuver definition. The utilized signal is either a sine chirp signal, which is a sine signal with increasing frequency over time, or a steering wheel angle course derived from a measures PSD. Based on either the measured steering wheel angle or the steering wheel torque as an input signal and the steering rack position as the output signal, the transfer function with gain and phase is calculated by the Fast-Fourier-Transformation (FFT) [45]. For feedback evaluation, the rack force is selected as the input signal while steering wheel angle, torque, or rack position function as the output of the system. An FFT analysis leads to the transfer function. Similar experiments were already performed in [5,24,25], although no correlation analysis was provided. For the rack force, either a chirp signal, or a pseudo-random-binary-sequence is implemented. The latter signal is well suited for system identification and possesses white noise characteristics, without the presence of clearly separated frequencies, as it occurs for the chirp signal [46].

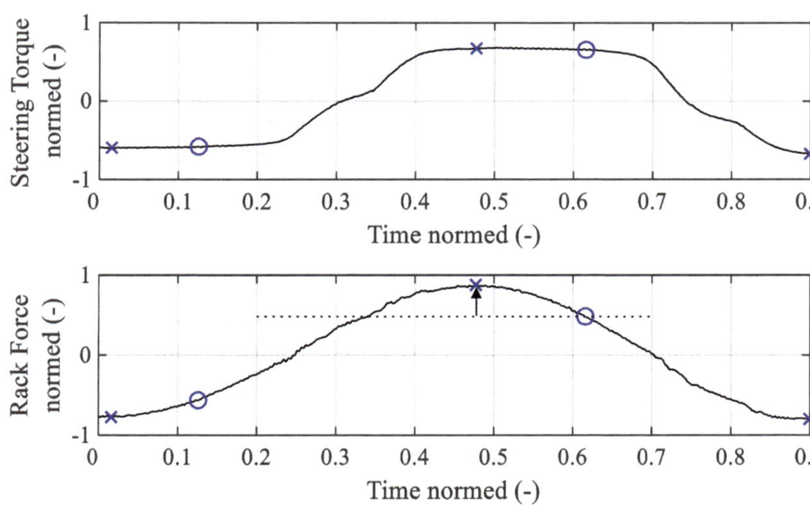

Fig. 11: Visualization of the rack force reduction value (black arrow)

In layer four, friction force and friction torque values are evaluated by implementing prescribed rack velocities or steering wheel velocities and measuring the necessary force or torque, respectively, to maintain the movement, see investigation from [5,25]. Additionally, the feedback transfer function is selected for evaluation. This also incorporates a new analysis for low frequency excitations during transfer function identification. The principle is displayed in Fig. 11.

The utilized metric is the rack force reduction, between the maximum rack force, see cross in Fig. 11, and the rack force at the timepoint when the measured steering torque is reducing again, see circle in Fig. 11. The black arrow indicates the rack force reduction value. Mechanically, there occurs a stiction effect since the rack force excitation displaces the rack relative to the steering wheel angle. When the maximum rack force is reached, the rack sticks in this position leading to a constant steering torque to hold the steering wheel angle in position, although the external excitation force is reducing. The black arrow indicates the rack force reduction after which the preload state is ending, and the steering torque reduces again. Since these stiction effects negatively affect the movement of the steering system if preloaded Is applied, a large force drop off is expected to correlate with negative friction perception.

The prerequisite for all correlation analysis here is to demand a logical connection between the objective value and the subjective criteria. This approach considers the critical aspects mentioned in [47], namely the "blind" connection of objective metrics with subjective assessments without practical meaning.

Out of the thiry-one described steering variants, 2 hardware variants are not considered for correlation since the assessment rating was too low and would deteriorate the investigated results. In these cases, the yoke was tightened as strong as possible, increasing the friction levels. Additionally, one software variant is neglected since the required signal for this function is not available on the steering test bench.

To eliminate outliers, all subsets containing 25 steering variants out of the residual 28 were regarded for correlation analysis. By doing so, there still exists an adequate number of subjective ratings while only 10 % of assessments are not incorporated for the evaluation.

For layer four assessments, only the hardware changes are considered for correlation analysis leading to a number of fourteen variants. Here, the same subset approach is implemented, incorporating fourteen out of fourteen variants.

5 Results

Since the Pearson coefficient is regarded as the metric to determine whether a subjective – objective connection can be identified, the threshold for relevance is set to 0.7, which coincides with the summary from [19] for test bench-based investigations. According to [48] this value is the lower limit for the identification of "high correlations".

5.1 Subjective Vehicle Evaluation

In total, as described in subsection 3.1, four subjective criteria under straight line driving and cornering are each evaluated, leading to a total number of eight aspects. In preparation for the test bench investigations, their interdependence, namely the correlation between different criteria is compared in Fig. 12 in form of the Pearson coefficient.

Subjective Criterion No.	1	2	3	4	5	6	7	8
1	1							
2	-0.7	1						
3	-0.08	0.08	1					
4	-0.12	0.03	0.65	1				
5	0.47	-0.57	-0.18	-0.22	1			
6	-0.72	0.79	-0.06	-0.15	-0.51	1		
7	-0.21	-0	0.74	0.62	-0.09	-0.04	1	
8	-0.3	0.03	0.46	0.58	0.02	-0.03	0.54	1

Subjective Criterion No.

Fig. 12: Pearson coefficient representation between the investigated subjective criteria from vehicle evaluations. Absolut values larger 0.7 are highlighted in black.

The threshold for an identified correlation is set to the magnitude of 0.7, identical like for the following investigations. As can be seen, only three criteria pairings appear to be correlated, namely friction impression during cornering with friction impression during straight line driving and response behavior. Additionally, steering feedback of single obstacles during cornering is correlating with single obstacle during straight driving.

5.2 Driving Simulator Testing

For the evaluation of the driving simulator scenario, the correlation is only considered for the same subjective criterion. Also, the influence of the visualization on a reduced subset of the subjective criteria is investigated.

Identified correlations

The correlation results between the driving simulator and the vehicle assessment are displayed in Table 4.

Table 4: Correlation results for the subjective assessment in the driving simulator to the vehicle assessment

Criterion	1	2	3	4	5	6	7	8
2D-Person (all)	-0.14	**0.90**	0.57	0.83	0.48	**0.95**	0.70	0.84
2D-Pearson (VR)	-	-	0.63	0.84	-	-	**0.79**	**0.92**
VR-Pearson (VR)	-	-	**0.95**	**0.93**	-	-	0.70	0.72

Identified correlations larger than 0.7 are indicated by a gray background, whereas the best identified correlation is printed in bold. For this chapter, the first row is relevant. Five out of eight driving simulator assessments correlate well with the vehicle assessment. Criteria "1" and "5" which incorporate vehicle reaction, namely response behavior in straight driving and cornering, do not reflect the assessment from the vehicle. Friction impression (criteria "2" and "6"), as well as feedback behavior display good correlations.

Therefore, a basic comparability of criteria which focus on the steering wheel interaction with the driver appears to be valid, while vehicle reaction influences are not well suited. It is important to state at this point, that both the heavy reduction of investigated steering variants and the utilization of solely boundary variants, benefit the Pearson correlation. Consequently, a larger study is required to support the herein derived results.

Influence of visualization

Comparison of the correlation results of the 2D monitor visualization to the VR glasses are also displayed in Table 4. Since the sample for VR evaluation is further reduced, the 2D-results are also decimated to the sample of the VR glasses. As indicated by the gray background, each feedback evaluation on the driving simulator exhibits a correlation to the subjective evaluation from the vehicle. Especially for straight driving with VR glasses, Pearson coefficients >0.9 are found, outperforming the results from 2D visualization. For cornering, this tendency is not observed, and the 2D approach outperforms the VR approach.

One explanation for this phenomenon is the feedback by the evaluator, who experienced pronounced nausea. During cornering, there is more movement within the visualization, worsening the effects for the driver when wearing the VR glasses.

Still, based on these findings the utilization of VR glasses may lead to improved correlation results when compared to 2D implementation. What needs to be considered, is that the VR evaluation followed the 2D evaluation. Due to the new task for the assessor, the gained experience during the first evaluations may have improved the performance of the assessor. Similar as in the previous subsections, future investigations need to incorporate these influences into the study design.

5.3 Objective Evaluation

In this subsection, the derived objective metrics are presented, the results from different abstraction levels are compared and a validation performed.

Identified Correlations

An overview of the best derived correlation coefficient for each criterion in each abstraction layer is displayed in Table 5 and the connected objective metric in Table 6. For layer 1, response behavior and friction impression are well represented by the gradient in steering torque over rack position and hysteresis values of the steering torque over steering wheel angle. Feedback evaluation displays good correlations for maximum occurring steering wheel angle excitations and rack position excitations. Similar results are derived for layer 2, which proves the good agreement between the measured steering wheel angle courses and the derived generic signals.

The identified correlation for rack displacement excitations during feedback evaluation also displays the benefit of the steering test bench since this measure is difficult

Table 5: Overview of the derived Pearson correlation coefficients for all criteria and abstraction layers for steering system alterations

	Layer 1	**Layer 2**	**Layer 3**	**Layer 4**
Straight, Response	-0.76	-0.75	+0.84	+0.90
Straight Friction	+0.80	+0.73	-0.82	0.82
Straight, Single Obstacle	+0.77	+0.79	+0.83	-0.72
Straight, Road Structure	+0.74	+0.72	+0.76	-0.86
Cornering, Response	-0.73	+0.00	+0.70	-0.7
Cornering Friction	+0.82	+0.77	-0.87	0.84
Cornering, Single Obstacle	+0.85	+0.89	+0.89	-0.82
Cornering, Road Structure	+0.00	+0.00	+0.75	-0.85

to obtain in vehicle applications. Additionally, it is possible that the change in rack position also occurs in the vehicle, which leads to reaction forces in the tire-road contact, which excites the vehicle. Although these vehicle excitations, which are perceivable by the driver, cannot be investigated on the test bench, they may be present in this objective measure.

For criterion "8", namely the friction impression during cornering, both layers are not able to deliver objective metrics with a Pearson coefficient > 0.7.

Whereas layer 1 and layer 2 provide accessible parameters, layer 3 parameters originate from frequency domain observations entirely. The gain and phase for each frequency bin of 0.1 Hz is saved and investigated for potential correlations. To avoid identification of false correlations, the Pearson coefficient is plotted over the frequency range, see Fig. 13.

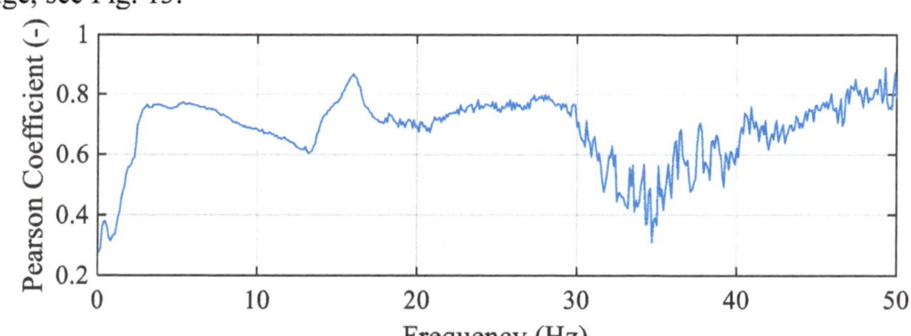

Fig. 13:Exemplary Pearson coefficient over the frequency range

Correlations for these metrics are only accepted valid, if a smooth course of the Pearson coefficient is visible in these plots, indicating a clear tendency of high correlation for this frequency range. Otherwise, it is discarded. For criteria which describe the evaluation of the forward path experiments for forward transfer function estimation are

Table 6: List of derived objective parameters for different layers

	Layer 1	Layer 2	Layer 3	Layer 4
1	Gradient steering torque over rack position	Minimum hysteresis steering torque over rack position	Phase steering wheel angle to rack position at 1 Hz (PSD)	Gain rack force to steering torque at 21.7 Hz (chirp)
2	Hysteresis steering torque over steering wheel angle	Maximum hysteresis steering torque over rack position	Phase steering wheel angle to rack position at 2 Hz (PSD)	Rack force reduction at 0.2 Hz (chirp)
3	Median of steering wheel excitation amplitudes	Average of rack position excitation amplitudes	Gain rack force to steering torque at 33.6 Hz (chirp)	Maximum fluctuation of rack force during friction measurement (0.5 mm/s)
4	Maximum of steering wheel excitation amplitudes	Median of rack position excitation amplitudes	Gain rack force to rack position at 16s.4 Hz (chirp)	Rack force reduction at 1.3 Hz (chirp)
5	Hysteresis steering wheel angle over steering torque	-	Phase steering wheel angle to rack position at 3.9	Minimum friction torque value (5 °/s)
6	Surface of hysteresis steering torque over rack position	Hysteresis steering torque over rack position	Phase steering wheel torque to rack position at 0.9 Hz (PSD)	Maximum fluctuation of rack force during friction measurement (0.5 mm/s)
7	Average of rack position excitation amplitudes	Median of steering torque excitation amplitudes	Gain rack force to steering wheel torque at 49.3 Hz (chirp)	Median friction torque value (1 °/s)
8	-	-	Gain rack force to steering wheel angle at 2.7 Hz (chirp)	Minimum friction force value (0.1 mm/s)

selected for correlation, while feedback criteria utilize feedback transfer fuctions, as explained before.

Based on the derived correlating gains and phases, incorporating transfer functions from rack force to rack position, to steering wheel torque and steering wheel angle, three frequency ranges can be derived, which are relevant for the investigated criteria, namely 0.2 Hz-2 Hz, 20 Hz and >40 Hz. The latter frequency range represents an interesting finding, especially when considering steering feedback. Although the steering feedback definition in this contribution does not include frequency ranges, typical approaches, as summarized in [6], set the frequency limit for useful information between 7 and 30 Hz, which was implemented as the maximum excitation frequency in different transfer function estimation techniques [5,25]. For the subjective criterion "7", a high correlation is detected for the gain from rack force to steering wheel angle at 49.3 Hz. Therefore, an indication is provided, that previous frequency limits for feedback evaluation are too low and useful information may include frequencies up to 50 Hz.

On layer four, the most generic and abstract layer, additional to the transfer function estimates, valid correlations are found for friction torque and friction force values. Good agreement to subjective evaluation is found for the rack force reduction value,

which is explained in Fig. 11. For friction force and friction torque, good correlations are derived for low rack and steering angle velocities.

Influence of Abstraction

For the first two layers, with exception of criterion "5" where no valid correlation was found in layer 2, there exist only small differences in the derived Pearson coefficients. Still, a tendency can be observed, that layer 2 provides lower correlation results, underlining the importance of precise replication of the driving situation.

For layers three and four, where the strict adherence to the surrounding conditions of the driving situation is neglected, improved Pearson correlations are derived, and there exist valid correlations for criteria "8" which was not found for the other two layers.

It needs to be stated, that the objective metrics in layer three and four represent the best correlation result, which possessed a logical connection to the subjective criteria. Since there is no agreement on a specific frequency range for parameter estimation and neither for a specific displacement velocity for the friction measurements, standardization of these boundary conditions may deteriorate the Pearson coefficient. Still, these results clearly indicate the possibility to deliver meaningful implications for the steering system performance on a very abstract level.

Validation Approach

To compare the correlation performance of each layer, the following Fig. displays a boxplot of the deviations resulting from vehicle assessment and the calculated assessment based on linear correlation models. The horizontal line in these plots represents the median value, while the upper box limit is the upper quartile, meaning that 75% of data lie below, and the lower box limit is the lower quartile, meaning the 25% of data points are located below this threshold. Blue circles indicate outliers which possess a deviation larger than the 1.5 interquartile range, meaning the distance between the box

limits. Residual maxima and minima which are no outliers are indicated by the black line [49].

For the comparison of the mechanical steering systems in Fig. 14, the median value is located close to zero for all criteria, indicating accurate estimation below the minimum rating step of 0.5. For layers one to three, the box plots are very similar, underlining relevance of the found correlations. Although the interquartile range for

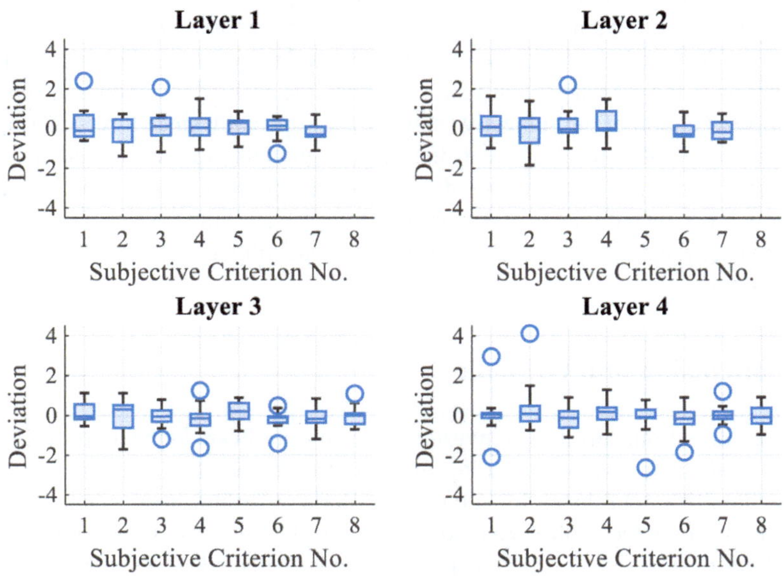

Fig. 14: Estimated subjective assessment for all layers on all mechanical steering system variations

layer four estimation is smaller compared to the residual layers, multiple outliers with more than two-point deviation occur which hinder its applicability.

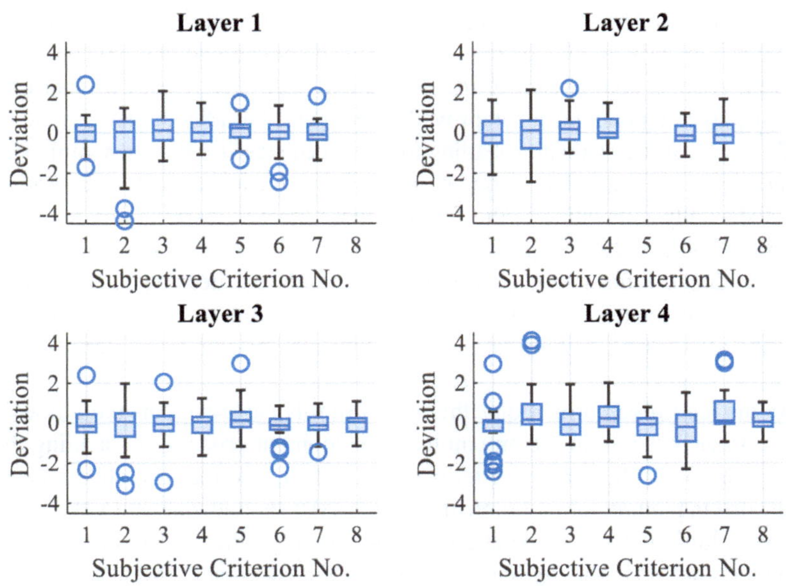

Fig. 15: Estimated subjective assessment for all layers on all steering system variations

If the subjective grades are estimated for all steering variations, the boxplot diagram in Fig. 15 is derived. Again, median values are approximately zero, indicating high performance. Although the maximum non-outliers and the interquartile range increases in comparison to the mechanical only investigation, the derived linear models represent a valid approximation of the subjective assessment, especially since three steering variants, which correlated the least with the derived models, are considered here, too. Based on these results, neither layer displays clear dominance in comparison to the others, which is especially interesting for layer four, because the passive characteristics of the steering system also correlate with the different software variants with the same hardware setup.

6 Conclusion and Outlook

In this contribution, correlation investigations on a hardware in the loop steering test bench are conducted, to assess steering feel outside the vehicle. Based on a literature review, a classification scheme for subjective criteria regarding steering feel is derived and objective parameters with already identified correlations are clustered into four categories. Additionally, a 4-Layer-Test specification is presented, allowing the investigation of subjective criteria on a test bench with different abstraction levels, ranging from ideal reproduction of vehicle test to the evaluations of characteristics of the steering system in a passive state. To generate a subjective reference assessment for the correlation analysis, thirty-one steering system variations with fifteen software alterations and sixteen hardware alterations, are subjectively assessed by on evaluator with eight subjective criteria in a real vehicle. Initial correlation investigations incorporated the use of a steering test bench which is extended to a driving simulator, to assess the variants directly subjectively outside the vehicle. It is demonstrated that valid correlations are derived. Also, a comparison to the assessment with VR glasses instead of a 2D monitor displayed a potential for improved assessment. Disadvantageous is the occurrence of nausea for the evaluator, leading to a reduced steering system sample size.

For the objective evaluation, strong correlations with each subjective criterion were derived. Especially interesting is the finding, that objective correlations for feedback assessment are obtained at 50 Hz. This frequency lays above previously reported values of useful feedback information. A validation, wherein the derived linear correlation models are implemented to generate subjective assessment for the mechanically changes and all steering system variants, displays the validity of the derived connections.

Due to the identified correlations, the results of this paper support the utilization of steering test benches within the automotive development process, by tackling the objective description of the subjective term "steering feel." Further investigations are required for driving simulator assessment to extend the sample size and validate its applicability. Additionally, considerations of the steering wheel fixation by the driver should be incorporated to increase realism on the test bench and find new correlations.

References

1. Deuschl, M.: Gestaltung eines Prüffelds für die Fahrwerksentwicklung unter Berücksichtigung der virtuellen Produktentwicklung. Ph.D. Dissertation; Technische Universität München, München (2006).
2. Harrer, M.; Pfeffer, P.E.: Steering Handbook, Springer International Publishing: Cham, Switzerland (2017).
3. Ersoy, M.; Gies, S.: Fahrwerkhandbuch, Springer Fachmedien Wiesbaden: Wiesbaden (2017).
4. Wolf, H.J.: Ergonomische Untersuchung des Lenkgefühls an Personenkraftwagen. Ph.D. Dissertation; Technische Universität München, München (2009).
5. Düsterloh, D.: Funktionsoptimierung und Komplexitätsbeherrschung im Entwicklungsprozess mechatronischer Fahrwerksysteme am Beispiel elektromechanischer Lenksysteme. Ph.D. Dissertation; Universität Duisburg-Essen, Duisburg (2018).
6. Grau, J.; Nippold, C.; Bossdorf-Zimmer, B.; Küçükay, F.; Henze, R.: Objective Evaluation of Steering Rack Force Behaviour and Identification of Feedback Information. SAE Int. J. Passeng. Cars - Mech. Syst., 9 (3), 1279–1304 **(2016).**
7. Heißing, B.; Brandl, H.J.: Subjektive Beurteilung des Fahrverhaltens. 1. Ed., Vogel: Würzburg (2002).
8. International Organization for Standardization: ISO 13674-1: Road vehicles — Test method for the quantification of on-centre handling — Part 1: Weave test (2010 (13674-1).
9. International Organization for Standardization: ISO 13674-2: Road vehicles — Test method for the quantification of on-centre handling — Part 2: Transition test (2016 (13674-2).
10. Harrer, M.: Characterisation of steering feel. Ph.D. Dissertation; University of Bath, Bath (2007).
11. Norman, K.D.: Objective Evaluation of On-Center Handling Performance.SAE International Congress and Exposition. Proceedings., pp 1–14 (1984).
12. Farrer, D.G.: An Objecitve Measurement Technique for the Quantification of On-Center Handling Quality.SAE International Congress & Exhibition. Proceedings., pp 103–115 (1993).
13. International Organization for Standardization: ISO 3888-2: Passenger cars — Test track for a severe lane-change manoeuvre — Part 2: Obstacle avoidance (2011 (3888-2).
14. Decker, M.: Zur Beurteilung der Querdynamik von Personenkraftwagen. Ph.D. Dissertation; Technische Universität München, München (2009).
15. Weir, D.H.; DiMarco, R.J.: Correlation and Evaluation of Driver/Vehicle Directional Handling Data.1978 Automotive Engineering Congress and Exposition. Proceedings. (1978).
16. Mimuro, T.: Four Parameter Evaluation Method of Lateral Transient Response.Passenger Car Conference & Exposition. Proceedings. (1990).
17. Fernandes, C.G.; Passos, L.; Mello, K.P. de: A New Approach to Improve Steering Response Development Using Bode Diagrams.SAE Brasil 2007 Congress and Exhibition. Proceedings. (2007).

18. Harnett, P.: Objective Methods for the Assessment of Passenger Car Steering Quality. Ph.D. Dissertation; Cranfield Universtiy, Bedford (2002).
19. Grau, J.: Objektivierung des Rückmeldeverhaltens elektromechanischer Lenksysteme. Ph.D. Dissertation; Technische Universität Braunschweig, Braunschweig (2018).
20. Groll, M. von: Modifizierung von Nutz- und Störinformationen am Lenkrad durch elektromechanische Lenksysteme. Ph.D. Dissertation; Universität Duisburg-Essen, Duisburg (2006).
21. Neureder, U.: Untersuchungen zur Übertragung von Radkraftschwankungen auf die Lenkung von Pkw mit Federbeinvorderachse und Zahnstangenlenkung. Ph.D. Dissertation; Technische Universität Darmstadt, Darmstadt (2002).
22. Fritzsche, M.: Ein Beitrag zur Objektivierung des Sicherheitsempfindens mit Fahrermodellen bei Lenkstörungen. Ph.D. Dissertation; Universität Duisburg-Essen, Duisburg (2016).
23. Nippold, C.: Prüfstandsbasierte Lenkungsoptimierung im Mittenbereich. Ph.D. Dissertation; Technische Universität Braunschweig, Braunschweig (2017).
24. Haas, A.; Menze, G.; Sieberg, P.M.; Schramm, D.: An Objective Evaluation Approach for Safety-Relevant Steering Feedback on a Test Bench. Vehicles, 5 (4), 1727–1742 (**2023).**
25. Schimpf, R.: Charakterisierung von Lenksystemen mit Hilfe eines Lenksystemprüfstands. Ph.D. Dissertation; TU Wien, Wien (2016).
26. Haas, A.; Schrage, B.; Menze, G.; Sieberg, P.M.; Schramm, D.: Subjective and Objective Evaluation of a Steering Function on a Steering System Test Bench.Ame 2023 - Automotive Meets Electronics. Proceedings., pp 50–55 (2023).
27. Gómez, G.L.G.; Nybacka, M.; Bakker, E.; Drugge, L.: Correlations of subjective assessments and objective metrics for vehicle handling and steering: a walk through history. International Journal of Vehicle Design, 72 (1), 17–67 (**2016).**
28. Marusic, Z.; Gies, S.: Das Lenkgefühl - Merkmale der subjektiven und objektiven Beschreibung. In Subjektive Fahreindrücke sichtbar machen I, pp 67–84 (2000).
29. Gil Gómez, G.L.; Nybacka, M.; Bakker, E.; Drugge, L.: Objective metrics for vehicle handling and steering and their correlations with subjective assessments. International Journal of Automotive Technology, 17 (5), 777–794 (**2016).**
30. Chen, D.: Subjective and Objective Vehicle Handling Behaviour. Dissertation; The University of Leeds, Leeds (1997).
31. Koch, T.: Untersuchungen zum Lenkgefühl von Steer-by-Wire Lenksystemen. Dissertation; Technische Universität München, München (2010).
32. Matschinsky, W.: Radführung von Straßenfahrzeugen, Springer Berlin Heidelberg (2007).
33. King, R.P.; Crolla, D.A.; Ash, H.A.S.; Whitehead, J.: Identification of Subjective-Objective Vehicle Handling Links Using Neural Networks for the Foresight Vehicle.SAE 2002 World Congress & Exhibition. Proceedings. (2002).
34. Kunkel, D.T.; Leffert, R.: Objective Directional Response Testing.22nd FISITA Congress. Proceedings. (1988).
35. International Organization for Standardization: ISO 7401: Road vehicles — Lateral transient response test methods — Open-loop test methods (2011 (7401).

36. Henze, R.: Beurteilung von Fahrzeugen mit Hilfe eines Fahrermodells: Ph.D. Dissertation. Dissertation; Technische Universität Carolo-Wilhelmina zu Braunschweig, Braunschweig (2004).
37. Neukum, A.: Bewertung des Fahrverhaltens im Closed-Loop. In Subjektive Fahreindrücke sichtbar machen II, pp 1–20 (2002).
38. Uselmann, A.; Preising, E.; Schrage, B.; Düsterloh, D.: A Test of Character for Steering Systems. dSPACE Magazin (02), 24–31 **(2016).**
39. Haas, A.; Schrage, B.; Menze, G.; Sieberg, P.M.; Schramm, D.: Improvements on the Dynamical Behavior of a HiL-Steering System Test Bench. Actuators, 12 (5) **(2023).**
40. Haas, A.; Schrage, B.; Menze, G.; Sieberg, P.M.; Schramm, D.: Oscillation and Vibration Reduction Approaches on a HiL-Steering-Test-Bench. IEEE ASME Trans Mechatron, 1–11 **(2023).**
41. Schramm, D.; Hiller, M.; Bardini, R.: Modellbildung und Simulation der Dynamik von Kraftfahrzeugen. 3., aktualisierte und ergänzte Auflage, Springer Vieweg: Berlin, Heidelberg (2018).
42. Mitschke, M.: Dynamik der Kraftfahrzeuge. 5., überarb. u. erg. Aufl. 2014, Springer Vieweg: Wiesbaden (2014).
43. Pacejka, H.B.; Besselink, I.: Tire and vehicle dynamics. 3rd ed., Elsevier/BH: Amsterdam, Boston (2012).
44. The MathWorks, Inc.: fmincon. Available online: https://de.mathworks.com/help/optim/ug/fmincon.html (last accessed 27 April 2024).
45. Isermann, R.; Münchhof, M.: Identification of Dynamic Systems, Springer-Verlag: Berlin, Heidelberg (2011).
46. Weißbacher, J.: Automatische Reglerparametrierung von Servoantriebsachsen. Ph.D. Dissertation; Technische Universität Graz, Graz (2016).
47. Krüger, H.-P.; Neukum, A.: Bewertung von Handlingeigenschaften - zur methodischen und inhaltlichen Kritik des korrelativen Forschungsansatzes. In Kraftfahrzeugführung, pp 245–262 (2001).
48. Zöfel, P.: Statistik für Psychologen im Klartext, Pearson Studium: München (2003).
49. The MathWorks, Inc.: boxchart. Available online: https://de.mathworks.com/help/matlab/ref/boxchart.html#mw_e9da6873-07c2-42c3-9cce-266f01fd1dd3 (last accessed 29 April 2024).

Power Supply Requirements for a Steer-by-Wire system

General disclaimer: The information provided below represents the current status of the activities and is subject to change based on further developments.

Goetz-Philipp Wegner, Ford-Werke GmbH
Functional Safety Engineer for Chassis Controls
Spessartstraße 50735 Köln
mail: gwegner2@ford.com

Alexander Ein Waldt, Ford-Werke GmbH
Technical Expert Steering Systems,
Spessartstraße 50735 Köln,
mail: aeinwald@ford.com

Dennis Kurzweil, Ford-Werke GmbH
Research Engineer for Advanced Power Supply & Energy Management Systems
Spessartstraße 50735 Köln,
mail: dkurzwei@ford.com

P. E. Pfeffer (Ed.), *15th International Munich Chassis Symposium 2024*,
Proceedings, https://doi.org/10.1007/978-3-662-71064-7_14

Abstract

In Steer-by-Wire (SbW), the mechanical connection between the Steering Wheel and Road Wheel is not present anymore; therefore, the steering power must mimic this missing connection virtually only. This is also leading to enhanced steering power and power supply requirements. These requirements need to be developed in the context of normal vehicle operation but also for operation at degraded modes in case of a fault. Additionally, typical automotive environmental conditions are challenging for the power supply architecture.

For intact systems, the derivation of requirements includes typical use cases like parking and driving, hence peak power, slew rates and average power demand are key parameters.

Generally, safe power supply architectures for SbW require a redundant power supply. This includes performance requirements for a safe vehicle degradation after a fault, requirements for the period of continued driving after a fault and requirements for the minimal energy storage to reach the safe state on vehicle level.

Exemplarily, for Battery Electric Vehicles, a power supply specification landscape for a SbW system will be presented and future challenges are identified.

Keywords:
Steer-by-Wire, Power Supply, Availability, Safety, Degradation, Transition, Topology, Peak Power, Average Power, Slew Rate, Voltage Drop, Voltage Quality, Temperature Management

1 Introduction

Since in Steer-by-Wire (SbW), the mechanical connection between the Steering Wheel and Road Wheel is not present, the SbW system needs to substitute this mechanical linkage. In consequence, this leads to a risk of loss of steerability also in case of power supply faults. The power supply design must provide redundancy with ASIL D to ensure steering availability after a power supply fault.

In addition, SbW must substitute the driver's steering power and enables enhanced steering functionality like variable gear ratio with speed dependent adaption of the ratio between steering wheel and roadwheel. This high-power demand needs to be ensured for all typical automotive environmental conditions.

This technical paper discusses the challenges associated with the development of a power supply architecture, as well as various aspects related to power supply sizing.

2 Relevant use cases

The power supply design must support relevant performance aspects for driving within an intact system as well as safety related performance in case of failed power supply redundancy. In this paper, the power supply redundancy is achieved through the implementation of two independent power supply channels (channel 1, channel 2).

2.1 Relevant use cases for an intact system

The intact SbW system shall feature robust functionality tailored for various situations, including parking and dynamic driving use cases. Parking typically drives the highest peak power demand because coefficient of friction between tire and road surface is high at standstill and low vehicle speed. In conventional power assisted steering system, the maximum steering angular speed is contingent of the combined power applied by the driver's force and the steering assist force. For SbW systems, the driver's ability to apply rapid steering angular speeds is less constrained due to the absence of the mechanical connection to the road wheels.

Especially when a variable gear ratio (VGR) is applied, the power demand required to turn the road wheels increases significantly due to the combination of high steering loads and very high steering rack speeds.

The following figures shows the difference between a constant gear ratio (CGR) comparable to a conventional steering system and a VGR. Figure **Fig. 1** shows the difference of the steering wheel angle while steering from lock to lock with similar steering wheel speeds. Figure **Fig. 2** shows the actual current demand for the road wheel actuator following the driver's steering demand.

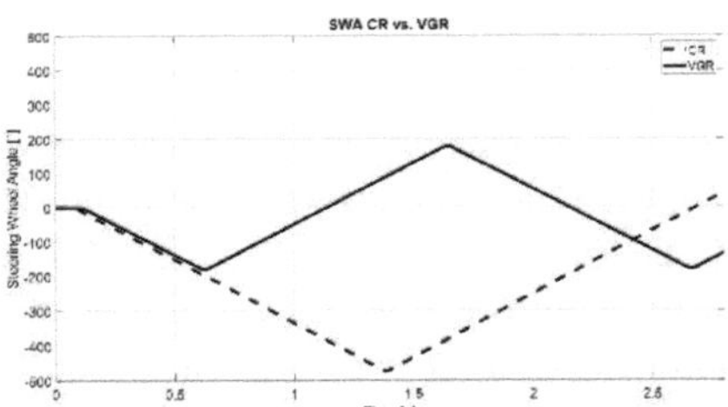

Fig. 1. Steering wheel angle lock to lock

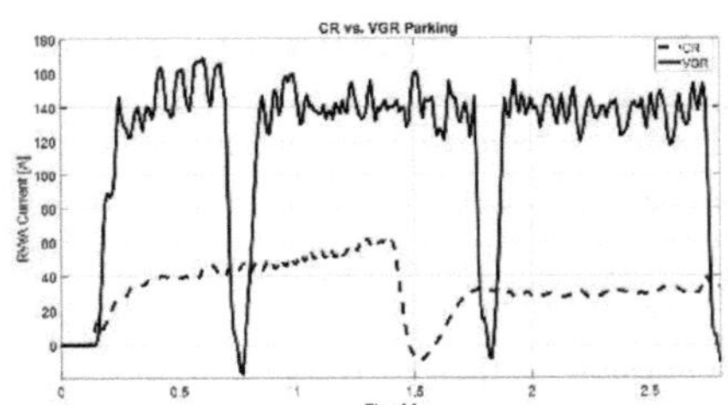

Fig. 2. Current draw for lock to lock

The measurement indicates that the power supply demand in case of a VGR is 2-3 times higher compared the CGR configuration.

If one of both, the power supply or the SbW system lack the capability to provide the necessary power for turning the road wheels, the road wheels will may not promptly respond to the driver's commands. This phenomenon, known as delay or misalignment, represents a significant concern as it deviates from the driver's expected steering response. The driver expects immediate and accurate vehicle maneuvering in line with their steering inputs.

In cases where inadequate power leads to recognizable delays, the SbW system must implement alternative measures to mitigate this issue. For instance, the driver's steering demand can be restricted by applying increased counter forces on the steering wheel, or compromises may be made during vehicle launch. These countermeasures may help to manage the system weaknesses and ensure that the vehicle's responsiveness aligns with the driver's expectations. However, those delays or their countermeasures have always the risk of decreasing customer satisfaction.

The other important use cases to size the power supply system are related to dynamic driving situations. This paper shows exemplary a double lane change maneuver according to ISO 3888-2 (**Fig. 3**). In general, the peak current, average current as well as current slew rates are lower as for the investigated parking situation with VGR (**Fig. 4**) before.

However, compared to the parking use case, it's essential to significantly limit the delay between the steering wheel angle input and the corresponding movement of the road wheels.

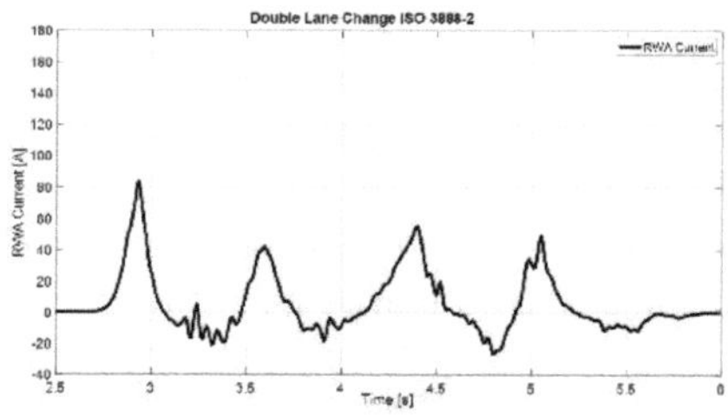

Fig. 3. ISO 3888-2 Lane Change

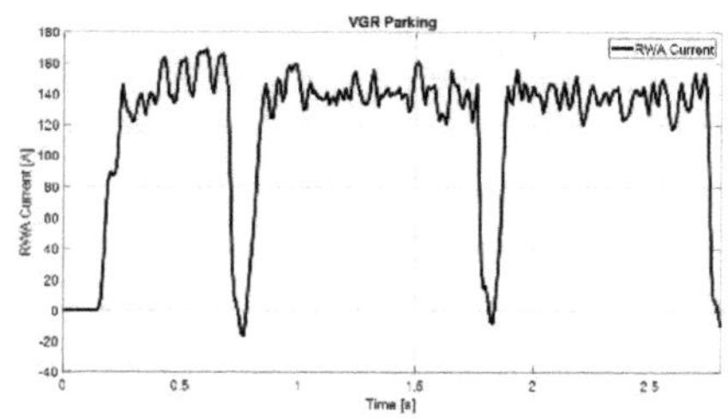

Fig. 4. Parking Maneuver (VGR)

2.2 Relevant use cases for a faulty power supply system

There are several aspects that impact the SbW system power demand for normal as well as for degraded operation in case of a power supply fault. The driver applied torque is no longer available and needs to be replaced by the road wheel actuator. In addition the feedback torque generation also requires some amount of energy. The peak power for feedback torque generation (simulating dynamic end-stop on steering wheel) and the peak power for accelerating the angular road wheel speed are typically not overlaid. That results in peak power mainly derived from the energy demand of the road wheel actuator.

Most peak power is required for accelerating and for achieving the angular road wheel speed at parking, which is equivalent to the energy demand.

A different group of important use cases addresses the minimal steering performance following a power supply fault, where one power supply channel is lost.

Both, the UNECE-R79 and the DIN 70065 specify vehicle level requirements for the minimum performance of the SbW system, as well as for the performance of the power supply system.

According to current UNECE-R79 specification, the remaining energy reservoir must be sufficiently sized so that the vehicle can complete 24 figure-of-8 maneuvers at 10 kph. This equates to approx. 35 minutes of driving time supported by the remaining power reservoir. The energy required for the SbW system to execute these figure-of-8 maneuvers represents a manageable amount, however, the total amount of required energy can by significantly increased by other vehicle components sharing the same energy reservoir.

Within the UNECE-R79 community, there is an ongoing effort to introduce an alternative approach for the dimensioning of the power supply reservoir. This approach involves decelerating the vehicle to 10 kph or bringing it to a standstill under certain conditions when a power supply fault results in the loss of the power source.

This alternative approach relies on advanced technologies to manage and monitor electrical storage systems, such as batteries. With this alternative approach, the size of the energy reservoir ensures that the vehicle is decelerated to 10 kph or lower. Typically, this duration is much shorter compared to the approximately 35 minutes as described above. Consequently, that leads to an optimization potential for the required sizing of the energy reservoir.

In addition to the energy reservoir, it must be ensured that the power supply can provide sufficient peak power to steer the vehicle. The required steering performance to execute specific maneuvers while driving at speed remain on the performance level for a conventional steering system. These UNECE-R79 maneuvers do not demand high vehicle dynamics, hence the peak power is predominantly restricted to low-speed situations where tire friction is substantial (**Fig. 5**).

In comparison, the DIN 70065 defines a minimum capability of dynamic maneuvers (**Fig. 6**) during transitioning the vehicle to the safe state. This includes minimum performance requirements of $4 m/s^2$ lateral acceleration for the specific maneuvers as well as the required energy reservoir for transitioning to a safe vehicle state. Additionally, the DIN 70065 provides guidance on time limitation for operating the vehicle at higher speeds, if only one power supply channel remains.

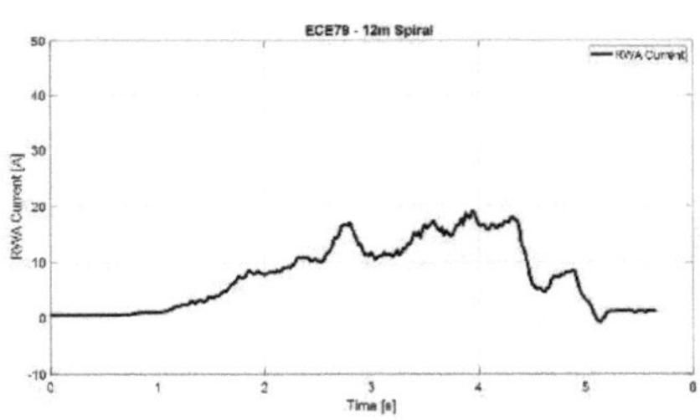

Fig. 5. ECE 79 12m spiral with 10kph

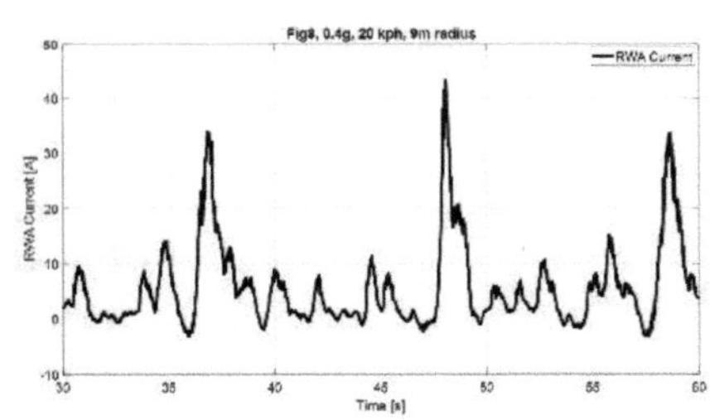

Fig. 6. DIN 70065 - Figure 8

3 Power Supply Sizing

The required power for a SbW system can be subdivided into the categories, peak power, average power and slew rate. Those properties need to be determined for each of the relevant use cases.

3.1 Comparison between EPAS and SbW

The required power for an Electric Power Assisted Steering (EPAS) or a SbW system is inherently dependent on the specific vehicle and its design characteristics. Factors such as rack forces (including axle load, return forces and friction) vary from vehicle to vehicle, as well as the efficiency of the steering system.

It can be assumed that a SbW system requires more power for the same vehicle compared to an EPAS system. This is based on the absence of driver power, a more direct steering ratio and the need of a second actuator.

The **Table 1** shows a comparison between EPAS and SbW (with VGR) for the same vehicle with respect to the peak power consumption.

Table 1. Example of peak power EPAS vs. SbW

	Parking	Double Lane Change (ISO 3888-2)
SbW system	~2000 W	~1300 W
EPAS	~1000 W	~1000 W

3.2 Variance of use cases

Developing a complete spectrum of test cases is indeed beneficial for designing the required power supply sizing points. Typically, there's a is broad range for peak power measurement results depending on test variance including the test driver. For instance, more dynamic turning behavior may lead to higher peak power requirements.

The following **Fig. 7** shows an example of peak power variances within the same test maneuver:

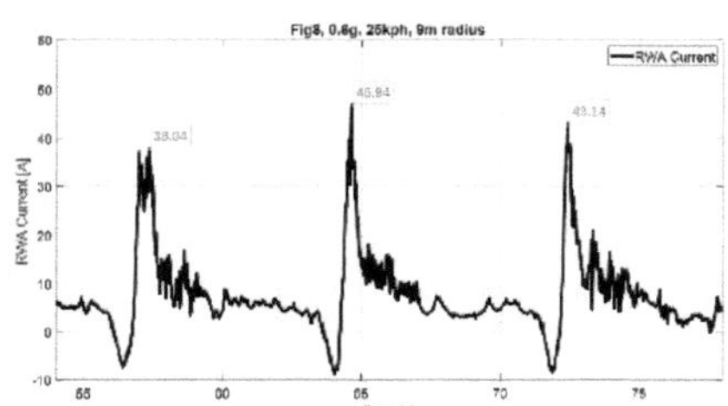

Fig. 7. Example of Peak Power Variance

3.3 SbW Design Challenges for high peak power

SbW challenges the power supply design by its high dynamic load characteristics. The power demand for normal steering situations is relatively low, but it can increase significantly during high-power-demand scenarios such as evasive maneuvers or lock-to-lock steering during parking. Therefore the power supply must be designed for the occurrence of high peak loads in combination with rapid rate changes (high slew rate).

This challenge is particularly pronounced within the regular 12V power supply network of current vehicles. Additionally, SbW actuators have limitations on maximum input current (limitation based on actuator as well as on wiring), further complicating power supply sizing. These actuator limitations (**Table 2**) restrict the allowable voltage range more as in conventional power supply designs. Therefore typically the inclusion of capacitive elements within the power supply design is required to ensure a constant voltage level under high transient loads.

Due to the limits of the SbW actuator, any voltage drop could result in insufficient power, potentially leading to steering delays. This is particularly critical in high dynamic driving situations where a significant steering delay can lead to safety-critical path deviations. Therefore it is important to mitigate voltage fluctuations in the power supply system caused by high dynamic SbW loads.

Table 2. Max current of steering actuators

	Typical max. current per channel
SbW (Road Wheel Actuator)	~80-120A
EPAS	~60-80A

3.4 SbW Flyback Current Challenges

Similarly to EPAS, Steer by Wire also generates high dynamic flyback currents caused by the kinetic energy of the vehicle. These flyback currents must be absorbed to prevent increased voltages beyond the allowed voltage range within the power supply network. Proper management of these flyback currents must be considered in power supply designs for all power supply states, including instances of power supply faults and their subsequent degradations. The following **Fig. 8** shows an example for flyback current in a highly dynamic maneuver:

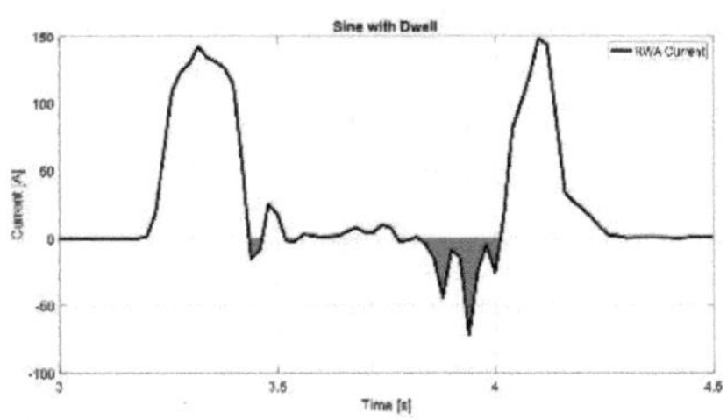

Fig. 8. Example of flyback current

4 Power Supply Architecture

4.1 Challenges for a SbW power supply design

The power demand of modern steering systems is significant, and it can be assumed that this power would continue to increase over the next years. A particularly challenging aspect is ensuring that sufficient power is reliably available for a reasonable duration under all circumstances, as specified. This necessitates increased requirements in the following categories:

Peak Power
Legacy steering systems typically draw between 60 A and 80 A per channel overall current (typically 60 A to 120 A for the complete EPAS system), with common system

voltages ranging from 12.5 V to 15.0 V. This results in a power demand exceeding 1 kW, with additional loads from other consumers of approx. 2 kW on top.

In the past this already led to sophisticated load shedding algorithms and usage scenarios. These aimed to manage customer demanded loads with the physically reasonable and affordable power sources. Typically, alternator and battery-based power supply systems offer a maximum power output ranging between 3 kW and 4 kW. Considering the above explained differences between EPAS and SbW systems and the particular power demands, it can be inferred that traditional power supply sizing is insufficient for SbW technology.

Voltage Quality

Due to the high peak power demand of the SbW system, the power supply must respond sufficient fast to deliver energy while maintaining voltage stability. This implies that the power sources must exhibit high dynamic behavior, e.g. an alternator typically requires up to 150 ms to regulate voltage setpoints. In Battery Electric Vehicles (BEV) or hybrid vehicles, the DCDC converter should be capable of controlling these setpoints in less than 5 ms. However, even with this rapid response, it may still not be dynamically sufficient. Therefore, batteries and/or capacitors are often integrated to further enhance the system's dynamic capabilities.

Voltage Drops

Examining the physical connection of sources and modules reveals another challenge: as the current increases significantly, voltage drops also increase. Depending on the sources and loads, cable diameters typically range up to 35 mm^2. For a 4 kW source such as a DCDC converter, an even higher cable diameter may be required to cover for power losses. Assuming a constant cable resistance, the power losses increase significantly in proportion to the amount of current flowing through the cable. However, the cable diameter and length are limited by weight and cost constraints.

Finally, a key aspect is the packaging of components to ensure minimal distances between power supply sources and high-power consumers.

Temperature Management

The commonly assumed automotive temperature range for power supply systems must typically cover -40°C to +130°C under the hood. BEV vehicles may offer the opportunity to reduce the upper range of the environmental temperature. However, this automotive temperature range has been a challenge in the past, but the criticality increases even more with the consideration of functional safety aspects of the SbW system.

In a BEV vehicle, the High Voltage (HV) battery and DCDC converter, although not capable of delivering full power at very low temperatures, it is still required that the vehicle can be started and driven. Another area of consideration is the low voltage power supply in BEV and hybrid vehicles. During normal operation without faults, sufficient power can be supplied via the DCDC converter. However, if the DCDC converter fails while driving, only energy storage systems such as batteries and capacitors are available to ensure a safe transition to the safe vehicle state.

Capacitors are limited in terms of the amount of energy they can store and may also experience a decrease in energy retention at low temperatures. In contrast, batteries can deliver high amounts of energy but are limited to provide the required power at low temperatures.

The figure **Fig. 9** illustrates a typical amount of power available from a battery when discharging for 1 second at the maximum current while maintaining a specific voltage level. The amount of available power is always significantly reduced at low temperatures.

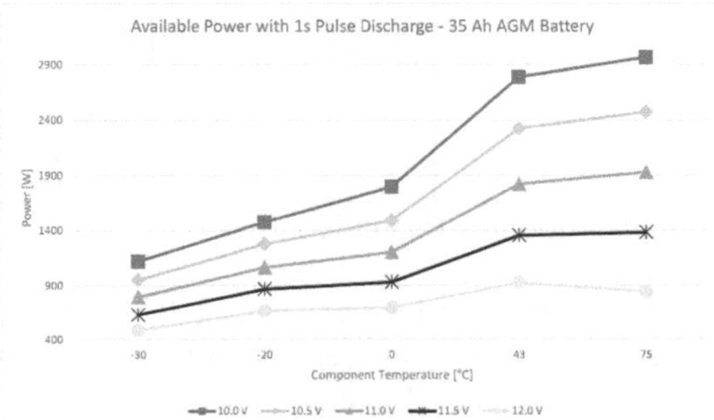

Fig. 9. Typical battery power (Ford internal data)

Transitioning the vehicle to the safe state

All of the above listed attributes support the safe operation of a SbW system. In the event of a power supply fault, it is essential to ensure that these attributes remain available until the vehicle is safely transitioned to a safe state (below 10 kph or standstill).

Therefore, independent redundant power supply designs are required to ensure that there is always a sufficient energy reserve along with power capability available in case of a power supply fault. In other words, relying solely on one power supply source without backup energy reservoir is insufficient to ensure a smooth transition to the safe vehicle state.

4.2 Power Supply Architectures

Considering all the challenges discussed in the previous chapters, three different power supply topologies will be examined. These are based on BEV and hybrid vehicle designs; nevertheless other power supply topologies are also possible.

A preferable architecture (**Fig. 10**) connects the SbW system with two DCDC converters and one energy storage unit. At least one low voltage channel is additionally supported by an independent energy storage device (such as a battery or capacitor) to provide steering availability after the loss of the HV traction battery.

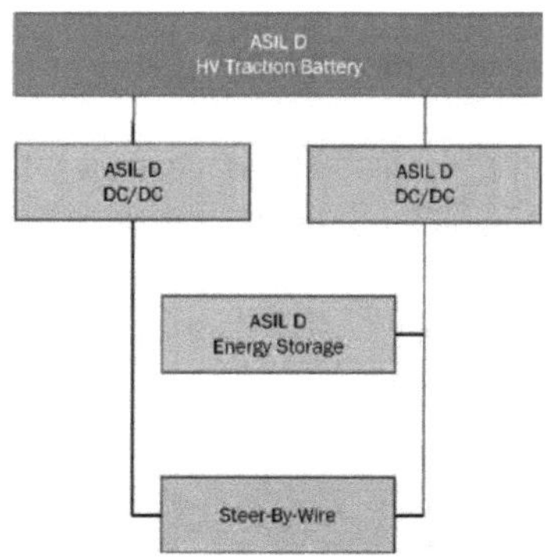

Peak Power	Limited by the DCDC converter sizing
Voltage Quality	Limited by DCDC dynamics. Can be improved by battery or capacitor elements on low voltage side
Voltage Drops	Packaging shall ensure optimized power distribution between modules.
Temperature Management	Thermal management is typically required for all batteries, potentially including energy storage on low voltage side
Transitioning to the safe state	Energy Storage is required to ensure transition to safe state if HV traction battery is lost. Energy storage allows typically short transition times.

Fig. 10. Power Supply Topology Example 1 BEV

An alternative architecture (**Fig. 11**) involves the supply of the SbW system via two DCDC converters connected either to different HV batteries or at least to two different independent battery packs within the HV battery. This setup ensures independent power supply sources for both SbW channels.

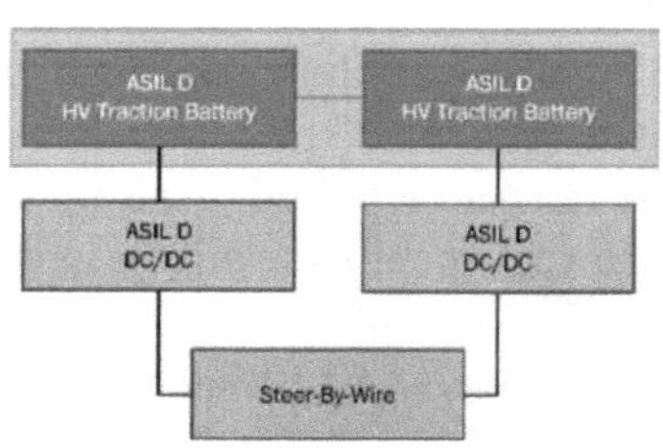

Peak Power	Limited by the DCDC converter sizing
Voltage Quality	Limited by DCDC dynamics. Can be improved by battery or capacitor elements on low voltage side
Voltage Drops	Packaging shall ensure optimized power distribution between modules.
Temperature Management	Thermal management is typically required for the HV traction battery
Transitioning to the safe state	Two independent HV sources provide redundancy. Longer transition times can be provided.

Fig. 11. Power Supply Topology Example 2 BEV

Additional alternatives for hybrid vehicles can include using both the generator and the HV traction battery as independent power supply sources (**Fig. 12**). However, due to the lower dynamics of the generator, an additional energy storage device is required.

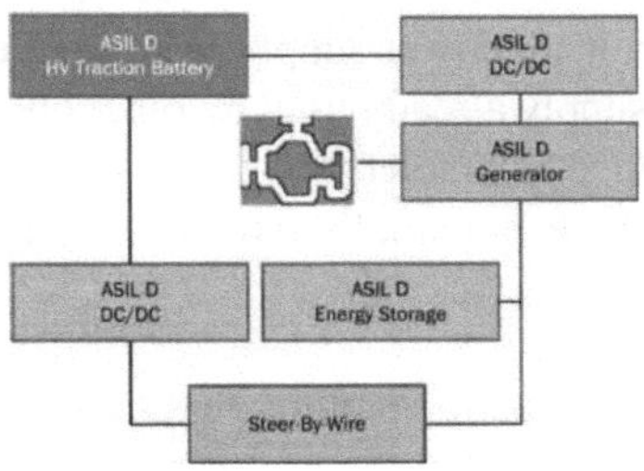

Peak Power	Limited by the DCDC converter and generator sizing
Voltage Drops	Mainly limited by generator dynamics must be combined with additional energy storage device(s)
Voltage Quality	Packaging shall ensure optimized power distribution between modules
Temperature Management	Thermal management is typically required for the HV traction battery
Transitioning to the safe state	Two independent sources by HV traction battery and generator provide redundancy. Longer transition times can be provided.

Fig. 12. Power Supply Topology Example Hybrid Vehicle

4.3 Power supply design for safety and non-safety critical consumers

It is preferred that the power supply of the DCDC converter is capable to supply non-safety-critical components as well **(Fig. 13)**. The parallel usage of safety and non-safety related consumers on one power supply channel requires a very dynamic separation of those consumers in case of a power supply related fault, e.g. to separate non-safety consumer if they short-circuit the power supply.

The separation of those consumers must be very fast, typically within less than 1 ms. Additionally, the SbW system also needs to resist these short voltage drops to prevent the resetting of the SbW Electronic Control Units.

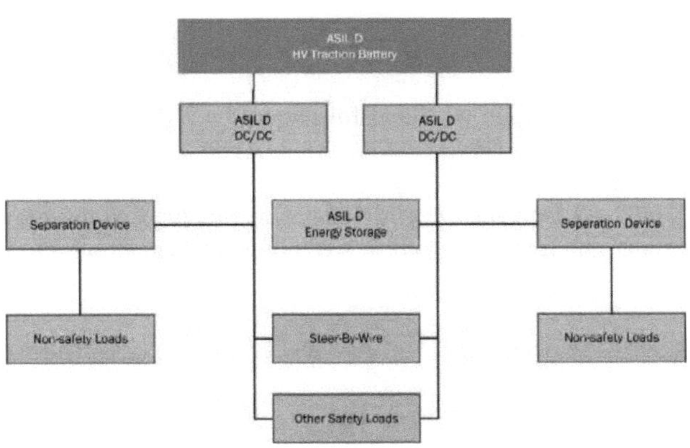

Fig. 13. Power supply topology example for safety and non-safety consumer

5 Outlook

The enhanced power supply requirements for SbW, especially the high peak loads, are significantly stressing the conventional 12 V low voltage power supply. Alternatively, voltage level can be increased, such as 48V power supply already used by other OEMS for SbW. This enables power supply cables with reduced diameters and limits the power losses due to lower peak currents. Higher voltage levels potentially reduce weight and cost of the power distribution systems.

Additionally, there is the opportunity to use the HV traction battery directly without the need for DC/DC converters. This approach would lead to the use of high voltage components outside the powertrain, and therefore introducing additional uncertainties associated with high voltage components. However, it potentially eliminates the need for special power supply components for the SbW system on the low voltage side.

UNECE-R79 and DIN 70065 specify requirements for the energy reservoir and power demand during the transition of the vehicle to the safe state. Those two standards enable a solid framework for the power supply sizing in case of faults. It will be interesting to see if the industry aligns on additional requirements for specific minimal risk maneuvers after a power supply fault.

Indeed, it will be fascinating to observe how the automotive industry future develops SbW systems and their corresponding power supply systems. The evolution of these technologies promises to enable significant advancements in safety, efficiency and performance.

References

10. International Organization for Standardization: ISO 26262:2018 Road Vehicles - Functional Safety (2018)

20. United Nations: Addendum 78: UN Regulation No. 79: Uniform provisions concerning the approval of vehicles with regard to steering equipment (Revision 5 – Amendment 4, 2024)

30. ISO 3888:2018, Passenger cars — Test track for a severe lane-change manoeuvre

Centralized Steering Functions for Software Defined Vehicles

Michael Story[1] Owen Tosh[1]

[1] Nexteer Automotive Corporation, Michigan, USA

Abstract. The automotive software development industry is undergoing a major shift as more and more OEMs are adopting a Software Defined Vehicle (SDV) architecture. Many high-tech companies, OEMs, suppliers, IT services, government agencies, contract houses, etc., have invested billions of dollars into SDV and features. As a result, automotive software is becoming more complex, and OEMs are looking at non-traditional suppliers such as startups and Silicon Valley as sources for new applications. This has led to a need for improved interaction between electronic control units (ECUs) and the increased use of domain controllers. Centralizing features such as EPS and Braking to a High-Performance Computer (HPC) in this new environment presents numerous new challenges and hurdles.

EPS/SbW software is an example of an existing actuator application that can be controlled from a domain controller in the vehicle. EPS/SbW software executed from a domain controller can provide the same level of performance as the same algorithm residing in the EPS/SbW actuator itself while allowing for greater interaction with other applications that may reside on the HPC. This presents challenges with lag times in communication and processing between the HPC and the EPS/SbW unit. Also, when systems are provided by different suppliers within vehicles, it becomes difficult to exchange information across ownership silos.

This paper takes a deeper dive into the challenges and potential solutions of providing an algorithm such as EPS/SbW on a domain controller in a SDV in the form of software as a product as well as solving these challenges while accounting for the OEMs' desire to execute software functions from an HPC. We present our own solutions in the context of what is currently being considered in the industry.

Keywords: Software Defined Vehicle, High Performance computer, Electric Power Steering, Stee by Wire, Communication Lag, Middleware, Domain Controller, Software as A Product, Standard interface, Driver safety, software marketplace, over the air purchase, over the air update

P. E. Pfeffer (Ed.), *15th International Munich Chassis Symposium 2024*,
Proceedings, https://doi.org/10.1007/978-3-662-71064-7_15

1 Introduction

The automotive industry is transforming from mechanical-centric systems with embedded software to a Software Defined Vehicle (SDV) architecture with centralized computing. The portability of the embedded software plays a vital role during this megatrend transformation. Autonomous driving, electrification, connectivity, security, safety, cost optimization, domain control, and cloud computation result in many software technical challenges. Original Equipment Manufacturers (OEMs) are inventing new SDV vehicle platforms based on software such that megatrend challenges and market business risks are minimized. Most OEMs are developing or will develop SDVs, some forming partnerships with high tech companies such as Google and Microsoft. Stellantis' "Stella Brain" [1] and GM's "Ultifi" [2] are examples of SDV architectures that use a centralized High-Performance Computer (HPC). According to Markets and Markets, the software-defined vehicle market is projected to grow from USD 270.9 billion in 2023 to USD 419.7 billion by 2028[3] as shown in Figure 1 below.

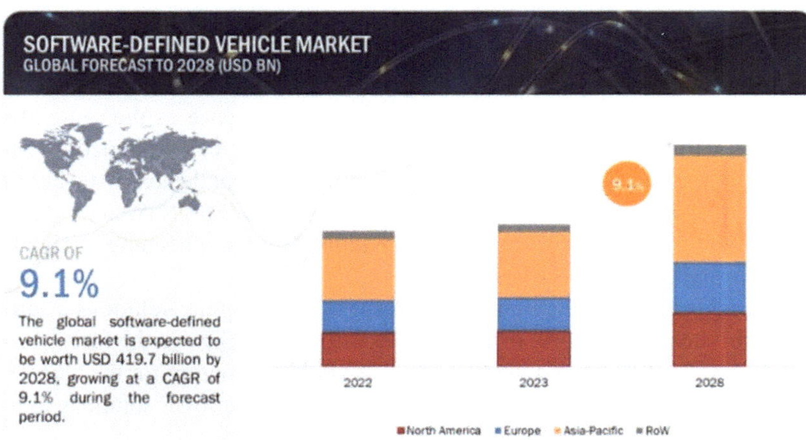

Figure 1: Software defined vehicle growth (MarketsandMarkets 2022)

The SDV platform supports updating and selling software after the sale of the vehicle. This enables a new business avenue for Tier 1 and software suppliers. A business model of selling software as a product to the OEMs has emerged. Software as a Product (SaaP) is an opportunity for OEMs to more quickly advance their software features offered to the customer, grow the supplier base for software, and update the software features after the sale of the vehicle.

Moving software functions from mechanical-centric system actuators, such as EPS or SbW, to a centralized controller allows data from many sensors to become available in the HPC, which require data from many different sensors and actuators for more complex algorithms. The complex software algorithm development leads to a need for improved interaction between the ECUs in the vehicle. This type of architecture brings many new challenges to traditional software development. The current vehicle system may have a controller with sensors connected directly to provide advanced driver-assistance systems (ADAS), such as advanced parking, lane-keeping, or advanced blind spot detection, using existing CAN messages such as vehicle speed, yaw rate, or other vehicle parameters. With a new HPC architecture, the sensor data must be retrieved on communication connections. Data latency, consistency, accuracy, and reliability play a vital role in the performance of the centralized HPC Steering features, whereas the local sensors directly connected to the actuator controller does not result in such challenges. Traditionally, a steering feature will be designed solely by one supplier where the supplier is responsible for the input, output, system response, driver feel, and safety of that feature. With new SDV HPC architectures, software can be provided by expert feature development companies, a feature developer must create tools to calibrate their features for the desired vehicle platform, along with the effective feature.

In addition, an HPC architecture is faced with challenges implementing software Automotive Safety Integrity Level (ASIL) development concerning the product development phase. There are mainly three phases of product development: prototype phase, production phase, and deployment phase.

Because software in the automotive field is evolving, it is difficult to find software feature classifications, interface standardizations, ASIL development processes, and understanding of latency implications. This paper focuses on such challenges for the centralized EPS features example. With standardization, tailored tool capability, security, operating system independence, and higher safety validation, EPS can be adopted for future vehicle architectures and support new software models. Even though every vehicle architecture is different, a software algorithm can be tuned for each vehicle platform and OEM's requirement. Since many suppliers are developing their unique software as a product (SaaP) solution, this paper will focus on EPS control.

2 Steering Functions

As an exercise of discovery, Nexteer Automotive created a proof-of-concept SDV architecture in a test vehicle. This architecture consists of an HPC controller, EPS controller, and four damping suspension controllers. Steering functions were moved from the local EPS controller to the centralized HPC. Steering functions in this case are those functions that provide assist, return, and steering feel. Determining which functions would reside in the HPC and which functions would remain on the EPS was an initial design decision in this exercise. While there are many places that line could be drawn, it was decided that all steering-specific functions would be moved to the HPC and the generic motor control functions would remain on the EPS actuator.

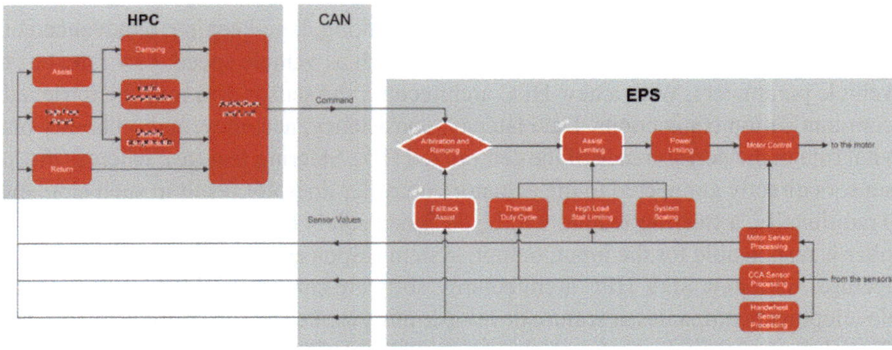

Figure 2: EPS Feature Separation in POC

In Figure 2, the block diagram shows at a high level the point of separation for the EPS software in the HPC. As can be seen the EPS at this point becomes a "dumb" actuator. The only functionality remaining on the EPS is motor control and processing of incoming sensor values. There is added limp-home functionality should the communication system fail. The communication in the POC system is dedicated 500KHz standard CAN. The CAN is a dedicated network shared between the EPS, HPC, and damper modules.

The EPS algorithm uses values that are native to the Steering system, Vehicle Chassis, and core vehicle systems. When an EPS algorithm is implemented in a Steering controller, concerns like data consistency, data latency, driver feel, vehicle control, data security, and vehicle safety issues are well understood and easy to overcome, but feature maturity and growth are limited. When an EPS algorithm is developed in a SDV HPC controller, there is more opportunity for growth, but the challenges mentioned are magnified and need to be addressed. Higher message rates, higher signal resolution, and redundancy could solve some issues, but data reliability and data standardization are more difficult to solve. Knowing this, why develop such EPS features in a centralized HPC? While centralizing functions like EPS presents challenges in implementing the existing EPS algorithm, it offers an opportunity to create new algorithms based on features that would otherwise be impossible to implement; data exchange, reliability, security, and data consistency between different ECUs' functions are improved. It also enables future feature updates more effectively. The industry learned during the chip crisis that minimizing and standardizing the chip suppliers helps vehicle manufacturers. The SDV HPC controller model enables users to combine multiple controllers into one bigger controller footprint.

3 Latency

When EPS features move from the local controller to the centralized HPC, performance parameters change due to latency and increased processing for data consistency. For example, a typical transmission rate for a signal on a vehicle network might be 10ms, 50ms, or 100ms, while the steering algorithm is processed every 2ms. Increasing the data rate and using a sample-average method to extract the required signal has been the traditional solution to latency, but this would require a protocol with a much higher data speed. In addition, there are levels of inconsistency in timing and protocol overhead, as well as safety concerns. Data coming from different protocols (for example, vehicle speed signal coming from CAN FD, and vehicle yaw rate signal from Ethernet) causes a data consistency challenge which may also affect performance. And because HPC controllers vary across different OEMs, it is difficult to design centralized EPS features to fit all.

The goal of SaaP is to provide the same software solution to all OEMs. Each OEM has its own vehicle architecture, message database, and cybersecurity. They also dictate what signals are sent and received to ECUs on the vehicle. Different suppliers to the OEMs have different interfaces to their specific ECU. These issues pose a challenge with multiple suppliers and OEM interfaces for a SaaP algorithm.

The EPS algorithm uses values that are native to the Steering system as well as other signals from the communication bus. When the EPS algorithms are moved out of the Steering system and into a centralized HPC controller, the Steering system now needs to send out all the sensor data needed by the EPS algorithms. The algorithms must wait for the Steering system to read, package, and send the sensor data before they can calculate an output and determine an appropriate motor command, introducing lag.

As Table 1 shows, the lag can be different with different communication types. This introduces an additional challenge: even if the challenge of executing the algorithms remotely can be overcome, they may run differently depending on the lag of the network. The accuracy of the algorithms may change depending on the communication lag, reducing consistency of results in different architectures.

Message processing time in μs					
User data bytes	CAN (500 kbit/s)	CAN FD (500 kbit/s// 500 kbit/s)	CAN FD (1 Mbit/s// 1 Mbit/s)	CAN FD (1 Mbit/s// 4 Mbit/s)	Ethernet (100 Mbit/s)
0	99.2	126.2	91.3	69.5	44.3
1	114.5	140.7	99.3	71.2	43.8
2	133.1	158.5	108.7	73.5	43.7
4	163.4	190.9	125.2	78	44
8	227.6	257.4	158.4	86.7	44
16	-	389.5	222.6	103.9	44.1
32	-	657.2	355.8	139.4	46.8
64	-	1171.4	611.8	208.3	53.4
1500	-	-	-	-	314.7

Table 1: Measured frame processing times for various transmission rates and user data size (Source: OTH Regensburg [4])

The solution is not just in increasing the communication speed. While this approach may help, it is not a definite solution to the time-based calculation problem. In any communication scheme, there is some level of inconsistency in timing.

The data available from each target vehicle for EPS is also different. This inconsistency makes it difficult for the centralized EPS supplier to provide a single solution for all OEMs. Currently, EPS must be reconfigured and recompiled for different vehicles. This approach adds to the cost, delivery time, and amount of testing required.

One other major issue is the communication delay that occurs with the addition of cyber security functionality such as AUTOSAR E2E and security key processing. Since each OEM uses different security algorithms, we tested the concept by artificially injecting a delay in the Nexteer Automotive POC software between the point when CAN messages are received and the model logic was executed to simulate the security delay.

3.1 Latency Effect

From the POC experimentation, the effect of latency on loop times can be seen in the system performance of centralized EPS on an HPC. Loop time for the experiment is defined as the time from EPS sensor read until EPS motor command applied as shown in Figure 4. This includes:

- Sensor reads in EPS
- Message transmit from EPS
- Message received in HPC
- Processing in HPC
- Message transmit from HPC
- Message received in EPS
- Processing of Motor torque command in EPS

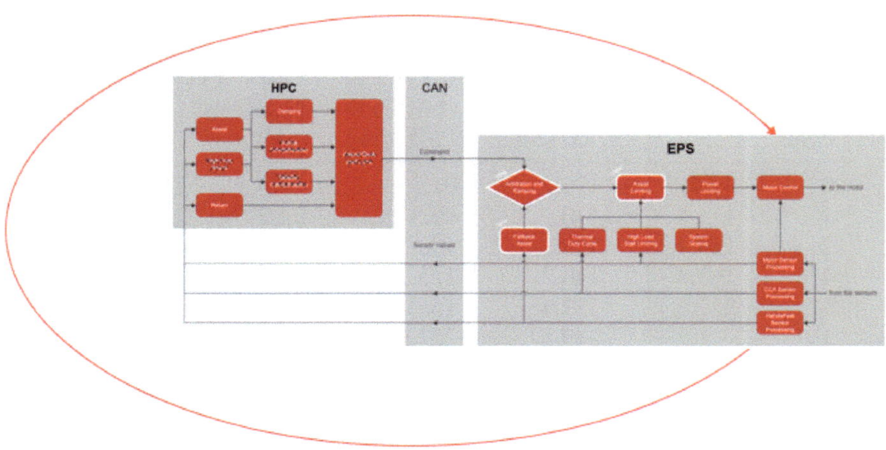

Figure 4: One loop, EPS sensor read until EPS motor command applied.

The perceived effect of these loop time delays is represented in Figure 5. As the time delay becomes greater the system performance in negatively affected. This behavior is very similar to what happens in a traditional EPS system when computation resources get overloaded. When the total loop time in system approaches 2mS, the steering feel begins to degrade, and instability becomes greater. Beyond 2mS, the system quickly becomes unstable to the point of being nonfunctional. The lag can come from anywhere in the system, processing, or communication delay. Cybersecurity can add time to both, demonstrating that increasing communication speed (such as by switching to ethernet) may not address the entire lag effect.

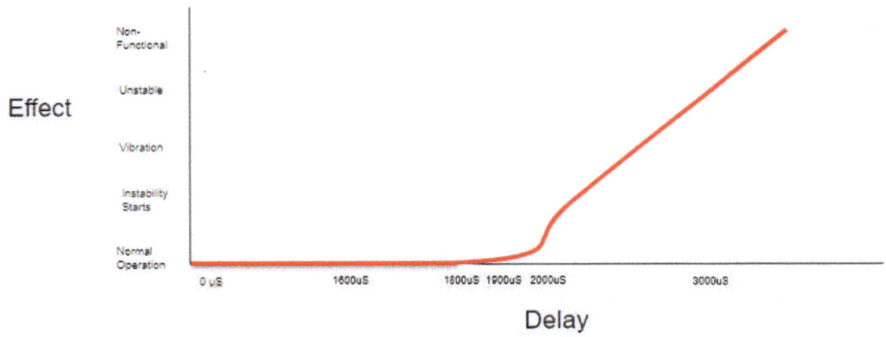

Figure 5: Effect of loop time delay in POC.

4 Communication Timing in POC

Communication timing between the HPC and the EPS controller is the critical bottleneck point for the system. The HPC EPS system must have the same steering feel and performance as the standard EPS system to be a sellable product. We can separate the EPS system into three main processing steps: sensor read time, torque calculation time, and torque application time. These steps execute on their respective processor with a message sent between each processing step. With respect to latency, all the processing and messaging must be completed in less than 2mS. Using a standard production EPS processor in our POC, processing times were measured as follows: sensor read time at 200uS, torque calculation time of 200uS, and torque application time of 60uS, for a total of 460uS of processing for the system between the HPC and EPS. The communication time for each processor is kept to a minimum with periodic message polling of 100uS and a transmission time of 150uS equaling a maximum of 250uS for each message, totaling 500uS for 2 messages. This makes the total round trip of processing and messaging approximately 1mS. This timing is well within an operating range to avoid any negative latency effects. The round-trip timing is illustrated in Figure 6.

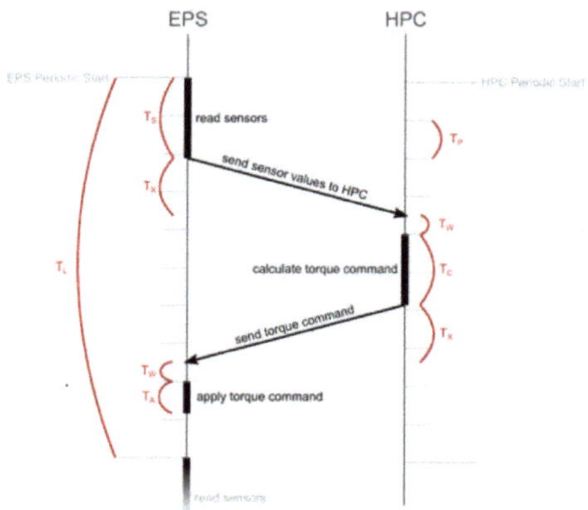

Symbol	Description	Value
T_L	Total loop time	$1ms$
T_P	Periodic (polling) time	$100\mu s$
T_S	Sensor read time (EPS)	$\sim 200\mu s$
T_C	Torque calculation time (HPC)	$\sim 200\mu s$
T_A	Torque application time (EPS)	$\sim 60\mu s$
T_X	Transmission time	$150\mu s$ @ $1Mbit$
T_W	Wait time	Max = T_P = $100\mu s$

Worst case execution time:

$$T_s + T_X + T_{Wmax} + T_C + T_X + T_{Wmax} + T_A = \boxed{906\mu s}$$

Figure 6: EPS HPC system timing.

4.1 Loop Time Frequency

In a system with EPS functions centralized on an HPC, one question that arises involves timing consistency. The worst-case timing can be calculated and tested. But the consistency of the loop timing can be a concern for systems that normally run locally on one microcontroller in a defined periodic rate. Figure 7 shows the occurrence of loop times in the POC system. This result shows the majority of the loop times being consistent and below 1mS with a small amount exceeding the average timing giving good performance. This has a positive effect on system noise and vibration that can be caused by more consistent loop timing. Filtering was also a factor in the differences in the loop timings in the POC.

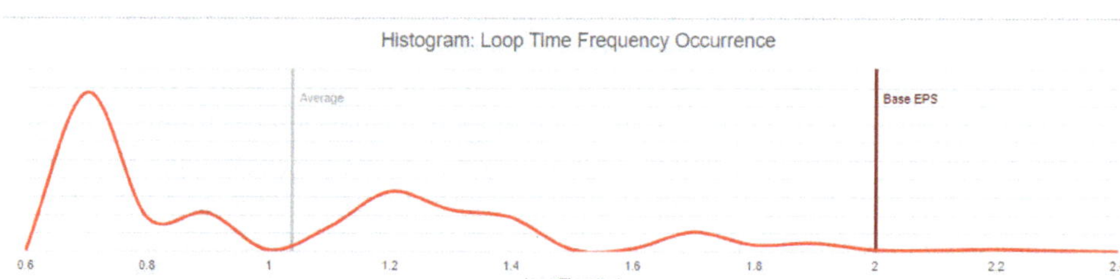

Figure 7: Loop timing occurrences.

The histogram shows:

- Average loop time: 1.04ms
- 99% of loops at or below 2ms
- Longest loop time: 2.4ms

The POC demonstrates that timing and consistency issues can be overcome so that the system has steering feel equivalent to a local EPS system.

5 ASIL & Quality

As OEMs move to a centralized HPC architecture, another challenge is introduced with software Automotive Safety Integrity Level (ASIL) development concerning the product development phase. There are generally three phases of product development: 1) prototype phase, 2) production phase, and 3) deployment phase. In the prototype / R&D phase, the problem statements are understood, ideas are generated, and prototypes are made for demonstration. In phase 2, the product is formalized, standardized, detail analyzed, and fully validated. Finally, in phase 3, the product deploys for continuous use, sustainability, maturity, and continuous improvement through a repetitive product life cycle. During each phase of development, questions arise about the appropriate ASIL rating of each feature. As shown in Figure 8, per ISO 26262 guidelines [5], software features can be designed at different ASIL levels.

Table 3 — Principles for software architectural design

	Methods	ASIL			
		A	B	C	D
1a	Hierarchical structure of software components	++	++	++	++
1b	Restricted size of software components[a]	++	++	++	++
1c	Restricted size and complexity of interfaces[a]	+	+	+	++
1d	High cohesion within each software component[b]	+	++	++	++
1e	Restricted coupling between software components[a, b, c]	+	++	++	++
1f	Appropriate scheduling properties	++	++	++	++
1g	Restricted use of interrupts[a, d]	+	+	+	++
1h	Dedicated management of shared resources[e]	+	++	++	++

[a] In methods 1b, 1c, 1e and 1g "restricted" means to minimize in balance with other design considerations.

[b] Methods 1d and 1e can, for example, be achieved by separation of concerns which refers to the ability to identify, encapsulate, and manipulate those parts of software that are relevant to a particular concept, goal, task, or purpose.

[c] Method 1e addresses the limitation of the external coupling of software components.

[d] Any interrupts used have to be priority-based.

[e] Applies for shared hardware resources as well as shared software resources, such resource management can be implemented in software or hardware

Table 8 — Design principles for software unit design and implementation

	Methods	ASIL			
		A	B	C	D
1a	One entry and one exit point in subprograms and functions[a]	++	++	++	++
1b	No dynamic objects or variables, or else online test during their creation[a, b]	+	++	++	++
1c	Initialization of variables	++	++	++	++
1d	No multiple use of variable names[a]	+	++	++	++
1e	Avoid global variables or else justify their usage[a]	+	+	++	++
1f	Limited use of pointers[a]	o	+	+	++
1g	No implicit type conversions[a, b]	+	++	++	++
1h	No hidden data flow or control flow[c]	+	++	++	++
1i	No unconditional jumps[a, b, c]	++	++	++	++
1j	No recursions	+	+	++	++

[a] Methods 1a, 1b, 1d, 1e, 1f, 1g and 1i may not be applicable for graphical modelling notations used in model-based development.

[b] Methods 1g and 1i are not applicable in assembler programming.

[c] Methods 1h and 1i reduce the potential for modelling data flow and control flow through jumps or global variables.

- **"++" indicates that the method is highly recommended for the identified ASIL;**
  "+" indicates that the method is recommended for the identified ASIL;
  "o" indicates that the method has no recommendation for or against its usage for the identified ASIL.

Figure 8: ISO 26262 Software development guidelines.

The centralized EPS software feature can be designed as a QM wrapper function [6] where the supplier would be responsible for most of the software while the OEM is reasonable for feature development. This arrangement brings additional challenges of validation and interface constraints.

The rapid software development methodology of Continuous Integration and Continuous Deployment (CI/CD) requires a software quality matrix to be considered at the beginning of a project. One approach to determining the software matrix is to define quality matrix criteria based on must-haves vs. wish-to-haves with the corresponding ASIL target by the software process quality team. For example, software repository, changelog, feature rationale, defect log, development toolchain, software feature time and space constraint, development platform detail and test confidence should be recommended in must-have at ASILD for centralized EPS from the beginning, while multiple compiler compatibility and library maturity fall in the wish-to-have list in the prototype phase of the development.

Functional safety aspects of the system would be handled with the same approach that is used for localized EPS. The responsibility would fall to the OEM or the centralized HPC developer for the HPC, and the EPS supplier for the local EPS. Close collaboration is needed between all software suppliers and OEMs to achieve the safety level required for an EPS system.

6 Software Interface Standardization

As software algorithm complexity and interfacing with multiple subsystems expands, the standardization of software interfaces is vital. A generic centralized EPS software solution that applies to all platforms and vehicles requires a more complex standardization interface. Software features developed with different ASIL ratings reside on the same HPC; I/O commands and MPU require a safety limiter firewall to cover severe error cases. Signal standardization is needed to make SaaP a more viable practice and force the industry to standardize the vehicle communication network. If an EPS system is sending out handwheel velocity, handwheel torque, and handwheel position for a specific OEM, it could be made standard for all OEMs. If they are not using the same offsets and resolutions, the EPS algorithm would have to change for each OEM. Standardization of signal attributes would significantly resolve many interface anomalies. This standardization would also help with the challenge of arbitration of the same signal value from different controllers in the vehicle. For example, different controllers in the vehicle use vehicle speed from different sources (wheel sensor-based, wheel speed-based, GPS-based). This would impact EPS performance. Having the algorithm's input changed from an absolute wheel speed sensor to a calculated value will affect the output of the algorithm. Organizations exist that are attempting to address the standardization issue, such as Scalable Open Architecture for Embedded Edge (SOAPHEE) [7] and The Connected Vehicle Systems Alliance (COVESA) [8].

7 Conclusion

The new trend of the software-defined vehicle platform brings new challenges and opportunities for software products. Software as a Product (SaaP) provides a new monetization model where many industries are constantly trying to develop their algorithms to tap into the new market segment. Implementing centralized EPS as a product on an HPC brings many challenges, such as standardizing interfaces, software development methodologies, software quality matrix, and software latency.

This paper considered challenges and potential solutions of providing an algorithm such as centralized EPS in the form of software as a product (SaaP) while understanding the OEMs' desire to source software functions from diverse suppliers, justify costs, and meet safety requirements. The complex software feature development leads to the need for improved interaction between the ECUs in the vehicle. For centralized EPS, this would address challenges such as data consistency, data latency, driver feel, vehicle control, data security, and vehicle safety issues. The Nexteer Automotive POC demonstrates the possibility of the implementation of centralized EPS on the HPC. The POC shows the issues of latency and consistency can be overcome with dedicated communication, increased bandwidths, and creative software implementation of message timing. Steering can also perform and provide equivalent steering feel to production systems.

The EPS example would support many vehicle platforms with interface standardization and common tool capability. With further development on security, operating system independence and higher safety validation the POC could become a production offering of centralized EPS.

References

1. https://www.stellantis.com/content/dam/stellantis-corporate/sustainability/csr-disclosure/stellantis/2023/Stellantis-2023-CSR-Report.pdf
2. https://news.gm.com/newsroom.detail.html/Pages/news/us/en/2021/sep/0929-ultifi.html
3. https://www.marketsandmarkets.com/Market-Reports/software-defined-vehicles-market-187205966.html
4. can-newsletter.org - Miscellaneous. (2022). Retrieved 30 April 2022, from https://can-newsletter.org/engineering/engineering-miscellaneous/210922_part-2-comparing-can-can-fd-and-ethernet_uni-regensburg_cnlm
5. ISO 26262-6:2018: Road vehicles — Functional safety — Part 6: Product development at the software level https://www.iso.org/standard/68388.html. page 23-32.
6. Keyur R Patel (2022). Model-Based Software Development – Autocode to Autosar. IETS 2022.
7. https://www.soafee.io/
8. https://covesa.global/

MIX
Papier aus verantwortungsvollen Quellen
Paper from responsible sources
FSC® C105338

If you have any concerns about our products,
you can contact us on
ProductSafety@springernature.com

In case Publisher is established outside the EU,
the EU authorized representative is:
Springer Nature Customer Service Center GmbH
Europaplatz 3, 69115 Heidelberg, Germany

Printed by Libri Plureos GmbH
in Hamburg, Germany